DEMOGRAPHY, TERRITORY & LAW: RULES OF ANIMAL AND HUMAN POPULATIONS

By S.M. Newman

BOOK ONE of FOUR in a series exploring the structure of population, economy and politics using new interpretations of evolutionary theory

Countershock Press

BY THE SAME AUTHOR

Sheila Newman, Demography, Territory and Law: Rules of Animal and Human Populations, Kindle Edition, December 2012.

Sheila Newman, *The Urge to Disperse*, Candobetter Press, 2011

Sheila Newman, Ed., *The Final Energy Crisis*, 2nd Edition, Pluto Press, UK, 2008

Andrew McKillop with Sheila Newman, Eds., *The Final Energy Crisis*, Pluto Press, UK, 2005.

Sheila Newman, *The Growth Lobby and its Absence in Australia and France*, Environmental Sociology Research thesis, Swinburne University, 2002.

Author's website: http://candobetter.net/SheilaNewman

Email: astridnova@gmail.com

Cover design based on Henri Rousseau (le douanier's) "Le Rêve".

DEMOGRAPHY, TERRITORY AND LAW: RULES OF ANIMAL AND HUMAN POPULATIONS

©S.M. NEWMAN
11 January 2013
5 March 2014

ISBN 978-1-291-17092-4

Countershock Press
PO Box 1173
Frankston
Victoria
Australia 3199

THE DEMOGRAPHY, TERRITORY AND LAW SERIES

This Book is the first of four volumes in the Demography, Territory and Law, which develops a new theory to explore the biological rules underlying modern politics, economy and demography and the future in a post petroleum era. The complete series consists of:

1. Demography, Territory and Law: Rules of animal and human populations

2. Demography, Territory and Law: Land Tenure and the Origins of Capitalism in Britain

3. Demography, Territory and Law: Land Tenure and the Origins of Modern Democracy in France

4. Demography, Territory and Law: After Napoleon: Incorporation of Land and People

The last three titles will be published in order and Book 2 is expected to be published in 2014. Book 3 should be available early 2015 and book 4 in 2016. Read more about the series at the end of this book.

TABLE OF CONTENTS

LIST OF FIGURES

FOREWORD: An Unexpected Solution to the Population Problem

The loss to infrastructure of biodiversity and green open spaces in rural and urban areas accompanying human population growth leaves many of us in despair. Conventional approaches to dealing with this problem of overpopulation universally rely on a counter-intuitive 'theory' of the '[benign] demographic transition' under a general paradigm of economic progress, which seems to mean more infrastructure and continual expansion, everywhere.

Sheila Newman, on the other hand, has, in *Book One, Demography, Territory and Law : Rules of animal and human populations*, revealed new solutions to these problems, which can be applied locally, voluntarily, almost automatically, if changes are made to planning and inheritance laws and if systems that currently function to maintain stability are preserved.

This means that population-problem experts need to look where they are not used to looking.

Newman's work, although initially informed by "Collapse theory" in the scientific tradition of Joseph Tainter, began with the question of why people in some societies and animals in some populations were able to live within their means, even though others failed to recognize the problem of overpopulation and overshoot of resources. In this her question also differed from a more prevalent one that goes, "Humans and all species seek to maximize their numbers, unless constrained by scarcity." Her answer is more than satisfying; it seems there really is a secret to successful stability, quite obvious when you know where to look. People have not looked.

Book One, *Demography, Territory and Law: The rules of animal and human populations*, is the first of four volumes, although it stands on its own. In clear, often humorous language, it establishes the need for a new approach and constructs and applies a new theory, bridging several fields. A description of the other volumes is located at the end of the book.

INTRODUCTION

The purpose of this introduction is to tell the reader something of what to expect from this book, how it is different from other books on similar subjects, why I wrote it, and why it is important. It is impossible to introduce the theory of *The Rules of Animal and Human Populations* without mentioning something of the way it will develop in the subsequent volumes of Demography, Territory and Law, because although it stands on its own, it foreshadows to some extent their subjects.[1]

The Rules of Animal and Human Populations establishes and explores a theory that there are distinct patterns of dispersal in all species, including humans and that, in human societies, these patterns affect the size of populations and the nature of economies and political systems. It looks at the rules common to traditional societies bound to particular localities, contrasting them broadly with modern settler states which have abstract concepts of membership and huge populations largely unrooted to locality and land. The other volumes in the series develop these ideas about complex societies and compare different land-tenure and inheritance systems and their effect on the economic and political systems in those complex societies, notably on the rise of capitalism in Britain and of democracy in France. The fourth volume will explore the reorganization of political and economic systems after Napoleon and the two world wars.

The Rules of Animal and Human Populations shows how different land tenure systems (or land-use planning and inheritance systems) may have radically different outcomes for human beings and the other creatures we share the planet with. It describes how one system can launch us into the insatiable demands of ever faster growth, dooming large numbers to overpopulation and poverty, whereas the other system can promote steady-state economies, equity and equality in a stable natural environment.

Wherever clan-based communities hang on to their traditional land-use and inheritance systems, people retain power over their lives and prevent their natural resources from being destroyed, exported, transformed and alienated.

A crucial difference between the traditional clan-based economies and modern ones is that the traditional clan-based systems do not buy and sell ('commodify') land.

A number of such traditional systems remain viable today – for example, in the islands of the Pacific. Some persist on small, rarely visited islands.[2]

Even today, in the face of aggressive attempts[3] to commercially contractualise these systems, in New Zealand and Fiji, indigenous land still cannot be sold in most cases; it can only be inherited.[4]

Intact societies remain in New Guinea, which is a large mountainous island with an extraordinary number and variety of different tribal peoples and natural ecologies. Although colonialism, transmigration, and globalism, have done and continue to inflict great damage to many of these peoples, there is evidence that their systems are self-repairing if the people are able to retain their control over land and the means of production.[5]

Royal and other 'noble' dynasties, although they may buy and sell land today, are nonetheless extremely conservative clan-based societies that function to preserve and enhance land and other wealth among closely related members, transmitting it via inheritance. Family corporations also function like dynasties. Both these models are significantly successful despite globalization and provide a distinct contrast with the dispersed and contractual nature of ordinary peoples' property and power in the vast settler states of the modern Anglophone world.

Importance and relevance of the theory explored in this book

The arguments in this book could be used to identify and validate the traditional long-lived systems in countries usually targeted for development and thus help to strengthen their peoples' solidarity and resolve to keep what they have and govern themselves.

Understanding how such extremely long-lived societies survive is also important to citizens of industrialized countries who are concerned about resource depletion, overpopulation and social, economic and environmental collapse at home. Identifying what land-tenure and inheritance patterns and laws can prevent such unraveling could be crucial to preventing or surviving collapse and to reestablishing order.

The different fertility outcomes between different land-tenure and inheritance systems are important because human expansion is placing the entire planet as well as our own species at increased risk of global climate change, fossil fuel depletion, water shortage, and desertification. While we expand, more and more of us suffer along with the natural environment. If we are really not able to stop expanding, we – in as much as the societies and landscapes that shelter us are - doomed. Even those who promote population growth as the road to riches do so mostly with the assumption that in the future population will 'transition' to stability. With some startling exceptions,[6] that is because they know it cannot go on indefinitely; it's just that they can't stand the idea of stopping growth now.

When complex society breaks down, we spontaneously reorganize into families and clans. When commercial society breaks down and money has no value, property ownership defaults to family connections with land, and nations default to families.[7] Some systems make this easy to do and others make it nearly impossible. This book identifies the qualities involved.

A criticism that has emerged from people casually perusing parts of my theory is that what I am describing belongs to the past and is therefore no longer relevant. This observation is wrong for several reasons. Modern dynasties, be they noble or corporate, continue to have strong effects on regional, national and international power and economies.[8] They are robust survivors in a disorganized world and their survival depends on their bloodlines. Clan and tribal rules also continue to define many modern complex societies, such as South Korea, which remains strongly affected by rigid kinship laws under its constitution despite industrialization.[9] Although Iceland[10] does not practice incest avoidance to the same degree as South Korea, urbanizing Icelanders have resorted to a special website in order to reduce the risk of accidentally incestuous marriage where confusion about lineage has been caused by movement away from traditional localities to the anonymity of the city. France has only moderate kinship-marriage restrictions[11] but the French retain strong clan links and habits connected to locality and a related underlying capacity to self-organise.[12] As I will show in my discussion of France in volume 3,[13] societies which have not been greatly disorganized in space tend to preserve underlying family and clan structures over long periods of time. These structures then present bulwarks and local organization points against top down principles of land-use planning and government.
A third reason is that when complex societies collapse, as I have mentioned, the default organising principle is back to family, clan, then neighbour. It helps if those principles of clan organization are already more or less intact in space.

The default organizing principle rule tends to undermine the strength of some 'peak oil' scenarios which assume that social breakdown following oil depletion will lead to gangs of men predating on women and rival settlements competing for power and resources by engaging furiously in population number competition – i.e. increasing their fertility. I think this supposition comes from erroneously assuming things can continue in the same competitive and uncooperative style as they have in large, fossil-fuel intensive societies. Although a period like this is sadly more likely in the highly diverse and commodified Anglophone settler states, such a period cannot last for long. Over time the family and clan system must reassert.

As I edit this book in 2012, many people would agree that international military activities in the Middle East are signs of polities and corporations forming blocs to compete for scarce resources. This kind of global power play also cannot last. In the absence of huge supplies of fossil fuel, local cooperation becomes inevitable due to the removal of distant connections and the limitation of resources to those locally present.

As you will read in this first volume of the four book series, *Demography, Territory and Law*, this default to family and clan carries intrinsic limitations to fertility, according to the balance of endogamy and exogamy. This means that in established societies with populations locally structured according to organic principles (i.e. those related to living organisms), overshoot[14] is unlikely to be a problem. Recognising this could make a huge difference to our freedom and basic prosperity. If we humans in the rapidly growing countries could protect and reestablish these largely instinctive principles of social organization against the global contractual social model, it is likely that we would stop helplessly destroying our environments and participate more effectively in democratic self-government.

Expansive economies and limited economies

Industrial capitalism is characterized by population growth, economic growth, expansion and intensification.[15] Although there was a short-lived case of industrial capitalism between about 1585 and 1622 in the Netherlands,[16] industrial capitalism as we know it today really originated in England, based on the apparently limitless fossil reserves of coal and the presence of many landless people who had to live by their labour.[17] These vast reservoirs of fuel gave rise to technologies that permitted mass production and facilitated the conquest and settlement of lands all over the globe. In the 20th century petroleum oil replaced coal as the fuel of industrial expansion, particularly where wheeled transport was concerned, although coal remained a mainstay for electricity production.

In contrast to Britain, America, Australia, Canada, with their larger landmasses, fuel reserves and well established trade links, small isolated islander societies had to learn to live within firmly fixed, immediate, territorial limits.

Any isolated island society which failed to do this must have perished quickly.

Two famous examples of reputed 'failure' to live within limits were Nauru and Easter Island (Rapanui). Nauru society had survived for many centuries, perhaps two or 3,000 years, but was quickly brought to its knees by capitalist 'success'. Remarkable for its natural wealth and fertility, it did not survive the transformation of its phosphate and limestone coral-island base into a guano-mine for global-based agriculture, despite access to large but transient commercial wealth.[18]

From a well-watered densely vegetated, undulating Eden of native gardens, it was razed within about 62 years (between 1906 and it independence in 1968) by mining shovels to a flat, waterless desert, within a narrow fringe of marginal sandy coast. Rainfall ceased to gather on the island, but simply ran off the edges. The once self-sufficient nation now imports almost all food and water from Australia.

Nauru sought compensation in the 1990s through the ineffectual International Court of Justice but finally settled with Australia outside the court for a paltry $107 million.

Nauruans are now a spectacularly unhealthy and divided people, who have been obliged to rent out their devastated country as an international refugee camp to the colonial Australian government.

Although industrial economic exploitation for outside interests quite clearly destroyed Nauru, the wider industrialized world knows little of this. The press portrays the tragedy of Nauru as something the islanders brought on themselves.

Whereas the fate of Nauru is accessible to any who care to seek – although few do - the decline of prosperous Easter Island (Rapanui) has become shrouded in mystery and myth. It has become the subject of famous inconclusive investigations and explanations, including Thor Heyerdahl's rafting experiment to test a theory that an earlier race of people built the famous giant statues on the island, and of Jared Diamond's *Collapse* book among many others. Diamond popularized a theory with elements of biblical parable, not unlike Sodom and Gomorrah, but minus God.

Such theories leave the wider industrialized world with the idea that the Rapanuians were ironically the masters of their own destruction.

It seems, however, that Easter Island was, after all, like so many islands, including beautiful, rich Haiti, just another island victim of piracy and colonial forces in the European Trade Wars. The overwhelming evidence for this will be presented further on in this book for Easter Island and in Book Three for Haiti.

Recognising the real causes of the destruction of Nauru, Easter Island, Haiti and similarly fated islands has far-reaching consequences.

For Jared Diamond, a North American, one reason to explore the failure to prosper of Easter Island and whole civilizations in the 'developing world' was to try to rationally dispel 20th century racist theories for persistent poverty and overpopulation among brown and black peoples. Another reason was because Diamond saw a parallel between the fate of Easter Island and the likely fate of capitalist growth economies, as doomed to collapse.

Diamond theorized that, like modern global citizens, the Easter Islanders failed to heed abundant signs that they were overpopulating and overexploiting their environment. For years this idea has been a popular subject of elaboration by ecologists like Tim Flannery in his book *The Future Eaters*[19] and bloggers and list-owners like Jay Hanson, who owned *The Dieoff List*[20] and pioneered *EnergyResources* among several famous resource depletion discussion lists. Just key Easter Island into an internet browser and you will find endless discussions about the mystery of the islanders eating themselves out of house and home as they continued to senselessly construct huge statues.

Ecological die-off writers' theorise that the major difference between Easter Island and nations in the global economy is one of scale. The apparent inability of Easter Islanders to take stock in time contained a dramatic irony due to the tininess of their territory, which meant that the warning signs should have been inescapable to all but the willfully blind. In these days of sophisticated modern communications and widespread education, similarities can easily be drawn.

Why, ecologists ask, do modern humans also sit like proverbial experimental frogs in slowly heating water, and fail to jump out of the saucepan or turn off the gas?

How is it that humans seem not to understand the incremental nature of systems failure, despite a myriad of gauges and dashboards, statistics and publications?

"Are we all doomed, like the Easter Islanders?" the cry goes out.

What if our fundamental assumptions are wrong and the Easter Islanders actually didn't befoul and destroy their own nest after all? I will argue a need to abandon the parable that the Easter Islanders brought about their own doom because it probably is no more true than that the Nauruans brought about theirs. (More about this later.)

The untrumpeted and inarguable truth is that Easter Island, like many, many successful Pacific Island societies, managed to survive for thousands of years. The durability of Pacific Island settlements is born out by DNA records which show that the *Lapita* people and their societies have been around for between 40 and 60,000 years, notably in New Guinea.

Major exploration and European colonization of Pacific Islands occurred a century or two later than in many other areas, like South America and Africa. The transition in the Pacific from self-sufficiency and order to overshoot and disorganisation over the 19th and 20th centuries was therefore in many cases carefully and scientifically documented by anthropologists whose excellent work now lies in dusty tomes. Not all of it caught the limelight of controversy like Margaret Mead's study of Samoa.[21]

If what emerges is that colonisation and capitalism destroyed these island cultures, does that mean that the same forces are now destroying the rest of the world as scientists like Jared Diamond and Tim Flannery seem to think?

Well, yes and no.

This book argues that there are two major systems in the industrial world. One of these systems is beyond the control of its citizens, but the other is not.

The work of Diamond, Flannery and others does not take significantly into account that the remorseless conditions of uncontrollable population growth and economic growth which are driving some major first world countries into overshoot do not really exist in the Western Continental European system.

Yes, Europe's population is too big to maintain without fossil fuel despite nuclear supplements already in place and it is causing unsustainable environmental destruction beyond and within its own borders, but the Europeans know this and they are preparing to 'downsize' thorough natural attrition. In fact population growth in Europe slowed dramatically from the first oil shock in 1972-73.[22] The problem is that the populations of North America, Britain, Australia, Canada, etc – the Anglophone states – failed to slow their growth after the first oil shock, even though this was apparently desired by some of their leaders.[23]

Not only did they fail to slow down, they have actually increased their population growth rate on bases which have more than doubled since the 1970s, in the face of increasingly serious threats to democracy, international politics and environment.

Since the oil price-rises and global financial shocks at the beginning of the 21st century, the Anglophone states have hugely ramped up aggressive exploitation of their own, as well as other countries' natural resources, with horrendous consequences to environment, democracy and international politics. (War with Iraq, Afghanistan, fracking for gas, blowing the tops off mountains for coal that had been considered uneconomic half a century prior, mining arctic national parks, etc.) These consequences have shocked and alarmed their own informed citizenry and the rest of the world. Yet the Anglophone states do not seem able to exert any more control over their fates than did the Easter Islanders, the Nauruans and the Haitians.

The problem of inertia in dealing with 'overshoot' in the Anglophone states is worsened by their institutional and commercial investment in continued population growth.

Europe's Roman law system seems to contain a solution in its retention of some features of traditional local societies. Starting with France, the system has been adopted by nearly every country in Europe since the 19th century, making the Anglophone system a minority one.

The principle difference between the systems is that Anglophone state citizens have very little democratic power to organize to defend themselves against a Market and destructive corporate interests to which they and their governments have become captive. In continental Europe, however, citizens and the state have quite effective power against corporate interests, largely because the state still controls most of the land-use. In the Anglophone states, private and corporate ownership of land, resources and utilities rivals that of the state.

Industrialization looks increasingly like a short-lived phenomenon. Unlike the successful Pacific Islander societies, industrial society has no experience of economies or polities lasting thousands of years. Despite this, business as usual in industrial societies, depends on a widespread assumption that industrial capitalist society will go on forever. Moreover, it is believed that the future holds material utopias which will gradually fuse into some kind of total knowledge. A popular book by Ray Kurzweil, *The Singularity*, is an excellent example of this new religion. Central to Kurzweil's thesis is an extrapolation of "Moore's Law" where trends based on past rate and production of technological improvement go on exponentially forever.[24] This works fine as an isolated theory, but the author gives little consideration to practical problems. The major practical problem is the concomitant trends in material, social and environmental costs[25] that underlie those past trends and which extrapolation of those trends would also increase.

Why did I write this book?

Someone had to write this book. Most books on population and environment don't straddle the human sciences and the natural sciences, but they need to. My tendency to read across disciplines and my initial research into what human institutions drove growth despite its democratic unpopularity caused me to stumble on some truths that did not strike others as important.[26] My ability to read in French and other Latin-based languages as well as English and to have done comparative research of social systems across countries was invaluable. An early degree in the history and philology of Roman language, which included the study of medieval history, helped me to realize that just because an idea is new, that doesn't make it better than others that went before. As a child spending time with indigenous Australian children in the desert regions of Australia during geophysical safaris, and later working for an Aboriginal organization that dealt with people from many different clans, I cannot help but see the validity of Pacific Islander cultures (which for me include Australian Indigenous ones) and the similarities with those of my distant ancestors who also lost their land. Apart from this I love more than anything the natural world and the freedom to enjoy it, the perfection of its highly organized diversity and the familiarity and comfort of its squeaks, squawks and songs, the wonder of its different faces, elements and skins. Continuous population growth is coercive to man and beast and I suffer personally from the loss of control over my environment and political system that it entails. The other values that influenced me were a love of science, the aesthetics of logic and the beauty of justice and of trees. How could I not write this book?

The reader may find it helpful to consult the very short Plan of this book which precedes Chapter One in order to get an initial at-a-glance grasp on the perspective and order of the arguments and concepts in this work.

Plan of this Book

In the first chapter we look critically at concepts of continuous economic growth, progress, and the notion of the 'demographic transition'. In the second chapter we look at modern theories of fertility, mortality and development. In the third chapter we look at 'land-tenure' or 'territorial' systems in other species, examining explanations for why they inhabit land in a more or less predictable fashion, and finding structural similarities between their territorial patterns and our land-tenure and inheritance traditions and laws. In the fourth chapter we formulate a completely new population theory to explain how humans and other creatures normally tend to live well within the carrying capacities of their environments and to allow space to fellow species. This is in contradiction to Hobbsian theory and much modern experience. In the fifth chapter, we look at clan-based land-tenure systems in steady-state Pacific Islander societies and how they fit into the preceding ecological theory, despite the apparent contradiction of the poster-child for short-termism, Easter Island. We then identify the things that can go wrong leading to modern social and political systems on continents and globally where population numbers, consumption and effective self-government seem to be beyond our control. We then foreshadow where our theory will take us in the three other books in this series, where we will find out why some industrialised countries are still relatively protected from overpopulation and have democratic governments, in contrast to those industrialized countries where populations and economic activity are growing far too big for their environments and democracy is shrinking.

CHAPTER 1: MAGIC AND THE SPACE AGE

Many people look forward uncritically to a future not too different from the world of the mid-twentieth century cartoon, *The Jetsons*,[27] where people have mechanical maids, whiz around in airborne cars, and take space-cruises to asteroids and far off planets. Politics continue benignly and poverty has no place. Political rank is preserved in the form of cheerful paternal talking head politicians on news-screens, but the slave and servile ranks have been transformed into machines which cannot feel pain or humiliation. A wealthy, not too intelligent middle-class of political spectators, has inherited this astro-playground and keeps order among itself with individuals switching roles between child and parent in an artificial recreation of the family and original clan system.[28]

The Concept of Progress

The explanation for this apparent chronological march to perfection is 'progress'. Belief in 'progress' is the 'modern' state religion, shared by communist and capitalist alike.[29] This is the religion that the West seeks to bring to the East and that the United States takes as its holy war against the Muslim defenders of Arab sovereignty over oil, often calling it 'democracy'. Some say it was Calvin who first unleashed the great God of progress, whereby the righteous were rewarded with power and wealth on earth. It was Darwin's thesis, but not Darwin's view, that was adapted by Spencer,[30] among others, to secularise this notion. The variables of population and resources and technology are constantly confused. A common interpretation of the demographic transition model of human population growth relies on the idea that population growth forces the invention of new technology, although it doesn't say why population grows in the first place and has no predictive ability.[31]

Progress, material growth and population growth: There is a space and volume aspect to progress. Progress demands huge amounts of materials and fuel for technology and mass production. As these materials and fuel run out in one locality the progressive economy and its population must expand to locate and liberate them wherever they can be found. The factories of progress demand many workers and their products require many markets. To a point yet to be located, the more workers available, the more materials and fuel can be liberated, the more factories can be built to produce products. Population growth is required for this expansion and expansion would be impossible and pointless without this population growth which creates more markets for the products of progress.

Progress and Ideology: This is why we have, on the one hand, an ideology that suggests that overpopulation is a bad thing and another one that suggests that drops in population growth are reasons to panic. The two attitudes have become hopelessly confused by the notion of the 'benign demographic transition' which suggests that for industrial societies you must first have overpopulation in order to have populations stabilise at some optimum level. The problem is that wherever population has been inclined to stabilise the priests of growth economics do everything they can to drive population growth up, by promoting higher birth-rates and immigration because economic growth depends on population growth to drive consumption and to multiply transactions.[32]

Progress and the nature of time: In the ideology of progress, time is relative only to human aspirations. Einstein notwithstanding, time is goal directed. According to this perspective, we humans face forward and march onward to perfection, every day getting better and better, continuously improving. We are taught to regard the past and old people with contempt because the further away from now, the further away from the future you are, the closer to imperfection, to ignorance, to naivety, to 'inefficiency', to 'primitiveness' or an earlier stage of 'development'.

Progress and destiny: Progress is not just an attractive option; we *must* have progress. We have no choice. Anyone who would stand in the way of progress stands in the way of wealth and human destiny and must be swept aside for ... for ...progress. So goes the circular argument.

Progress and Democracy: The mass media market 'Progress' and manage any little rebellions along the way. For instance, as human population growth drives competition for land it brings about ecological destruction and denies people access to familiar places and activities. These changes give rise to concerns over loss of sovereignty and outrage our sense of place. Our reactions to being boxed in and dictated to then tend to stick in the gears of the progress machine. These human limits to the machine of progress are a part of wider thermodynamic processes by which everything from landscape to metabolism simplifies and eventually loses its identity. Progress hastens this process of 'entropy'.

Progress and Profit: The modern mainstream media has evolved as the mouthpiece of corporations. Indeed it has become inseparable from them. It is owned by them and it owns them. It is a collection of corporations with interests in just about everything. It is a collection of seats of power. Media corporations do not just sell TV programs and newspapers. They own and sell property, mines, materials, natural resources, technologies etc. Increasingly they own governments because politicians and governments depend on the mainstream media to deliver their campaigns to the electorates. The corporatised media does not deliver campaigns for politicians that do not do what it wants or who wish to reform it. Politicians who condemn the progress ideology are characterised as kooks by the mainstream media.

This does not mean that they really are kooks, but perception is what matters and the media control perception.

Progress and coercion: As limits begin to impose themselves in many different ways on this principle of endless human expansion and populations groan with resentment at being manipulated to serve economies, ideologies and spins must be found to keep the mob moving, even in Pacific island former paradises. In early 21st century Australia, aggressive, self-styled 'no-nonsense' stances set the tone for coercion, as in this manic article for the Brisbane Courier Mail, Australia, entitled, 'Damn 'em all'.[33] In it the writer is talking about the Queensland State government's attempts to force a new dam on a region in order to cope with a growing population's increasing demand for water. That the same government invited interstate immigrants to the region and caused the problem in the first place is glossed over.

> "…. The sky is falling. The end of the world is upon us. Our cities are too big, choked with gridlocked traffic and toxic fumes. …We're running out of energy resources, the greenhouse effect will end up frying us all and as the temperature rises we won't even have enough water to drink. …Fix it, but please, not in my backyard. Or, as the new acronym BANANA, build absolutely nothing anywhere near anything. … Tough."

The writer acknowledges that there are scary problems. A stoic supporter of Progress, he doesn't protest at the costs; he volunteers for sacrifices.

> "Someone has to pay a price for progress. And I for one accept that living in a big, fast-growing city comes with noise, air pollution and, increasingly, higher-density living."

Living in a big city is posed as an unquestionably desirable thing. In this way any challenge to the idea that population growth itself might be halted is pushed aside. Anyone who would not desire to live in a densely populated city is crazy, ungrateful, or unrealistic. The costs of supplying water are trivialised; at no time does the writer canvas anything more than the near future. At no time does he question the constant additions to the population.

> "...Now we are ... looking at dams in catchment areas where occasionally water does still fall from the sky, such as the Mary River dam. Yes there will be communities hurt by this. Yes there might be a rare purple-striped eighteen-hyphen gilled trout ... inconvenienced by... extra megalitres of water washing around. If you don't like it, come up with a viable alternative."

By avoiding questioning the necessity or inevitability of population growth, the writer can use the arguments of conservationists against them by pretending that there is no choice. If there really were no choice then protest would indeed be unreasonable. He hints that more unpleasant decisions are in the wind: recycling of effluent, desalination plants, and nuclear power plants. We are expected to swallow a lot of s*** for Progress. The writer reviews the menu, as he sees it:

> "Recycling effluent? No, can't have that because, well, because it sounds yucky. Desalination plants? No, no, no ... can't have them, they use too much energy. Ah, energy. There's a touchy subject. The tree-huggers don't like coal-fired power because the gases the power stations emit allegedly will cause global warming. But wait, we can't have nuclear power because before you can say Chernobyl we'll all be glowing more brightly in the dark than one of the mutant cannibals from the hills have eyes."

He then ironizes the dangers wind turbines cause for birds and the problem that solar panels come from energy-intensive factories and would not adequately power a city.[34]

Progress and the "Green Movement": The writer is correct to say that many who identify as 'environmentalists' under a political brand of "Green movement" which is more advertising and ideology based than ecological science based, are energy, technology and population ignorant, believing that we can adjust to endless growth benignly. But the writer himself has no idea of the size of the problem. Most of his proposed solutions are not only finance and energy costly; their capacity is limited and their expiry dates are constantly brought forward with population growth. But we are assured that the Emperor of the time really does have a wardrobe of new clothes. Granted they will be costly, but the outcome will be splendid:

> "...Three cheers to Prime Minister John Howard then for truly opening up the debate about nuclear power. Yes, it does work. Yes, its emissions are nothing like carbon-based power, and yes, it's reliable."

In the same way that an internal combustion engine requires a car to be built around it and roads to run on, factories to build these, mines to find materials and economies of scale involving mass production, the nuclear power plant needs huge amounts of infrastructure, mines, chemicals, land, water and transport systems. Currently it uses those provided by fossil fuel – which is carbon-based. If we took fossil fuel out of the system and tried to replace it with nuclear power, the costs of a complete makeover of infrastructure, transport design and land-use planning would be even more important than the technicalities and energy costs of fuel supply.

The writer of 'Damn 'em all' seems to be unaware of the reliance on fossil fuel in current production of nuclear fuel. In fact the cost of building conventional nuclear plants and of processing nuclear material and managing the products and bi-products remains fossil-fuel expensive and pollutant. It has been asserted that every stage in the nuclear process, except fission, produces carbon dioxide.[35]

You cannot just mine (using trucks and other machines) then burn radioactive rocks in a furnace to get power and they don't produce enough radioactivity for our fuel needs in their natural form. You have to refine the rocks down to a much purer product that no longer resembles a rock. Getting ^{235}U out of uranium ore (yellowcake rock) is called the 'enrichment process' or 'separation process'. This process still requires electricity, even though the amount has been greatly reduced recently by using centrifuge instead of gaseous diffusion.[36]

There is a way of getting the radioactive material to create fission that involves a lesser concentration of uranium ore and that is to use it in combination with "heavy water". This has been for some time the established technology in Canada, home of CANDU reactors. These are 'heavy-water reactors'. They use a process of distillation[37] to separate heavy from ordinary water. 340,000 pounds of ordinary water is required to produce one pound of heavy water in the distillation process.[38] Indicative of the energy costs, the price of heavy water was around $300 per kg in 2008[39] and in 2012 it was about $712.[40] Another measure of the cost of getting heavy water is how much it costs as a proportion of original plant. This has been estimated at 20% of the capital cost for one CANDU reactor.[41]

Coal-fired electricity is used to keep nuclear plants cool, and petroleum generators are the fall-back when there are power blackouts. As the world realized with the 2011 Japan earthquakes and tsunami that affected Fukushima, the power plants cannot keep themselves cool when they break down. In Fukushima the emergency generator ran out of petroleum-based fuel and mains electricity were cut off.

We also hear about 'Gen IV reactors' – breeder reactors that create most of their fissile power by generating continuous fission from tiny amounts of material and reusing their own waste - but there are currently no such commercially functioning animals, although experimental ones do exist. Investment marketing exists to make it sound as if they are already up and running reliably ready to fuel entire nations with electricity.

'Damn 'em all' clearly believes that the new 'solutions' he favours will be as elegant as their alternatives will be ugly and atavistic, but closer attention to the problem shows that it is not that simple if you look past the propaganda.

"New transport corridors. New sources of energy and new water supplies. Go for it fellas, I'm happy my rates and taxes are going towards them. If you don't like it, leave. Go and find some drought ravaged shrubbery outside civilisation to live under, build a bicycle made of dead tree roots and make fire from the leftovers. ... ."[42]

Needless to say, there are few places to run to due to 'development' and there isn't much firewood, due to land-clearing. We are captives of this 'Progress'. We cannot easily get free. And the writer of "Dam 'em all" wants to dig us in even deeper.

Progress and the Demographic and Economic Transition: We might say then, if this is progress, do we really want it? And why do we have to have it? The final justification is the myth that progress brings wealth and if we continue to have progress all the poor in the world will ultimately be wealthy and this will cause their fertility rates to drop, due to the 'demographic transition'. So to deny progress its victims is to deny the poor economic justice and to condemn them to poverty and overpopulation. This is also the argument for first world population growth, open borders, and labour deregulation. It seems to suit the churches, big business, and land speculators too.

Progress and Propaganda - Life before the Industrial Revolution: But the argument relies on confining our appraisal to a short period.[43] It is true that things have improved for the first world in the 20th century. It is true things *were* worse 164 years ago in the 'first' world. In 1842 the British working class had an average life-expectancy of 17 years. 57% died before the age of five. The effects of coal on air quality and light were so devastating that the life expectancy of the city gentry was only 38 years and about half of the population of the industrial areas had rickets, which is a vitamin deficiency disease due to lack of sunlight.[44] (The rural poor lived to be around 38 and the rural gentry had an average life-span of 52). At the time of the Crimean War 42% of recruits from the cities and suburbs were rejected due to their poor health and small stature.[45] This misery was multiplied by the first ever globally explosive population growth that fed the industrial revolution and fed from it, which saw the English population increase from about 5 million in 1750 to 21.5 million in 1881.

But it is not true that life was always like that until the 20th century and that things have *never* been so good in the first world or the third world.

Progress and Measuring the quality of life: The quality of life is often measured by the average health, height and life-expectancy of populations. Demographic statistics can appear to bolster up life-expectancy claims that turn out to be spurious. Frequently cited demographic transition dogma about life expectancy ignores situations where many people died early and the impact of other events – such as pandemics - which knocked out large portions of populations. The Black Plague drastically reduced the *average* lifespan in Europe, without necessarily reducing the lifespan of those who survived it. Romans are often described as not very long-lived, but infanticide was common practice for much of Ancient Greek and Roman history and, more or less covertly, in all societies in different eras, some of them not so long ago. The very high mortality of the British working class[46] and the decline in birth rates during the 1890s depression reflect this practice as well as socially-linked neglect, malnutrition and disease.

In fact it seems likely that, before the industrial revolution, and before feudalism, many people lived long lives. "The effective end of the human life span under traditional conditions seems to be just after 70 years of age."[47] People also grew straight and tall.[48] Average heights of men in Europe fell from 173.4cm in the early Middle Ages to approximately 167cm during the industrial revolution. Return to earlier stature did not occur before the twentieth century.[49] Likewise the healthiest groups in the Americas predated Columbus's arrival.[50] Descriptions of Africans and Pacific Islanders, including Australian aborigines, by those who first encountered them give similar pictures. Nomadic herders like those in Afghanistan or the proverbial Arabian sheik (pre-petroleum era) also appeared as tall, straight and proud. Those who have retained land and continue to cling to their traditional lifestyles, whilst stressed by the constant battling entailed, also appear tall, straight and proud.

Wide scale poverty, social disorganisation and dispossession within societies seem to have occurred with the institutionalisation of feudal society. Feudal society imposed hugely inequitable land distribution, reducing previously free communities to the dependency and servitude of livestock. Families lived in cramped quarters with insufficient land-rights. As feudalism broke down, other events combined with a series of legally instituted land-enclosures to create permanent populations of landless people, especially in Britain.[51]

These landless people had nothing but their labour to sell and they became the 'working class'. First they laboured on farms. Later they provided factory fodder. Church and government policy embodied in the poor laws compensated landless women for having children and the men who married them. Child labour increased the earning capacity of landless families.[52]

The advent of the industrial revolution in Britain actually made things even worse than they had been made by agricultural feudalism, since it brought about an even greater series of enclosures and dispossessions, which arguably were carried over to colonies like Australia, Africa, India, the United States, New Guinea, various other Pacific islands, etc - in waves of immigration – and are still ongoing.

This later series of land enclosures began around 1750 in Britain bringing labour, coal and ironworks together; transforming the countryside. These enclosures responded to the structural demands of mass production and gave rise to the densely populated industrial slums around the coal-mines, iron-foundries and ports, where mining, manufacturing, building and exporting all went on at once, demanding huge supplies of labour. Involved in all these activities and dominating the new industrial towns were the first corporations. In these slums real life expectancy dropped to possibly its lowest ever level in Britain. It was these industrial slums that formed Engels' political theory and led to his association with Karl Marx.[53]

The lot of the working class improved only gradually on the lot of the medieval serf from the early 20th century. So, after the privations accompanying the feudal middle ages, there were about three centuries of absolute misery for larger and larger numbers of people. This misery was exacerbated by the explosive population growth that fed the industrial revolution and fed from it, which saw the English population increase from about 5 million in 1750 to 21.5 million in 1871.[54]

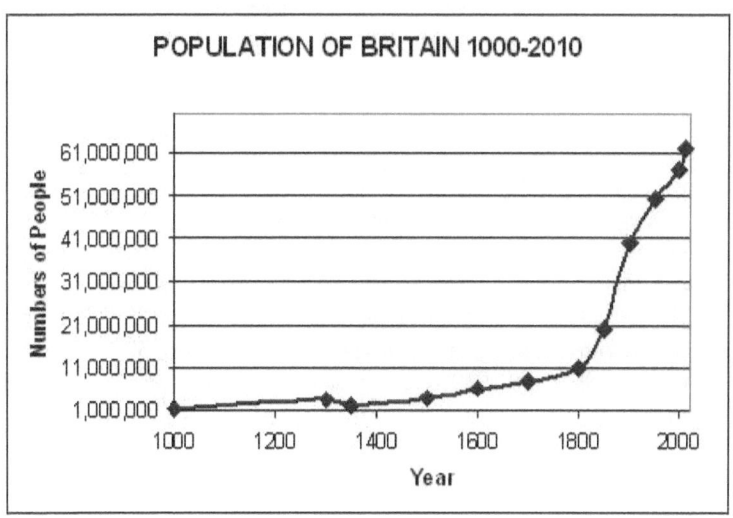

Figure 1. Population of Britain 1000-2010.
Source: Data are from national censuses from 1801 and UK Office for National Statistics. Earlier population estimates are highly speculative, from various sources including http://www.bbc.co.uk/history/british/launch_ani_population.shtml

No wonder Malthus became concerned about exponential growth. Already, by 1750, London was the biggest city in Europe.55 There was no comparable simultaneous population explosion in neighboring France or anywhere else, although something on a smaller scale happened in the Netherlands in the 16th century.56

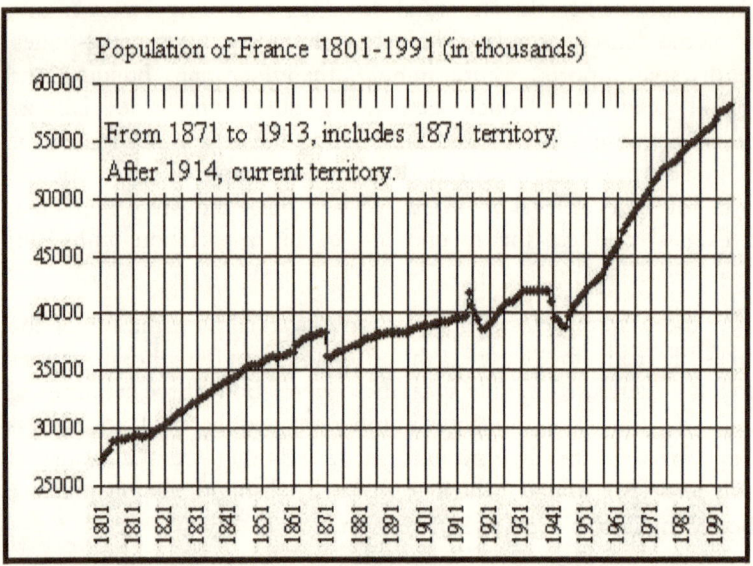

Figure 2. French Population 1801-1900.
Source: Francis Ronsin, *Histoire de la Population Française*, Seuil, Paris, 1998. Permission to use granted by author.

Figures 1 and 2 show population history graphs for Britain and France respectively. Although different in points of commencement, they are comparable where it matters. That is to say that the sudden acceleration in population growth and unprecedented rise in numbers in Britain from about 11 million in 1800 to about 46million in 1939, an increase of around 35 million, is obvious, whereas France grew from around 30 million in 1800 to about 43 million in 1939, a relatively gentle increase of about 13 million. Growth only really accelerated after World War 2. The period includes the addition of nearly one million people when Nice and Swiss Savoy became French in 1850 and a dip coinciding with the Franco-Prussian War in 1870-71 and the loss of territory in 1914.

The demographic transition theory relies on the expectation that every polity will experience similar circumstances to Britain's industrial revolution, based on assumptions that land-use planning and mineral wealth are dependent variables on human behaviour constants. The human behaviour constants are inferred from once-only's that accompanied colonisation. These were population explosions in new lands by the migrants there that were followed by falls in fertility in these colonial populations which can be related to the rising cost of land, fuel and other natural resources, which were forced up by the inflation that always accompanies population growth and finite resources where people have control over their lives.[57] But the demographic transition theory ignores rising costs in such cases and infers reproductive restraint in an overall positive economic climate.

Progress and 'externalities': The fates of the indigenous peoples whose lands were over-run and seized for a different economic purpose seem to be treated as 'externalities', in the same way that natural resources are. Yet, the promise of wealth and democracy for all is still used as the excuse for the dispossession by 'development' of entire peoples, leading to their enslavement, marginalisation or sometimes even to their extinction. But some modern economists whom the mainstream media has assigned the role that priests used to occupy in propping up the *status quo*[58] find the 'theory' of the demographic transition useful and don't appear to have any more capacity to look at it critically than priests where economics has displaced the pre-industrial religions.

Rapid population growth in developed countries and the demographic transition theory

An implicit assumption underlies the demographic transition theory. The assumption is that, after a period where industrialisation or 'development' coincides with excessive population growth, that growth will decline and then stabilise. However real observed populations have not borne out this assumption. Saudi Arabia, America, Australia, Canada, the United States and the United Kingdom continue to grow rapidly[59] whereas most parts of Europe are on course for a relatively abrupt decline in numbers around 2050,[60] after which they will probably stabilize down to pre-baby-boomer levels, or lower depending on fossil fuel and other depletions.[61]

Fear of Exponential Demographic Implosion

Some demographers have even expressed concern that failure to continue population growth may launch an exponential decline in human numbers leading perhaps to the extinction of homo sapiens,[62] which in the current context of 6.5 billion homo sapiens seems counter-intuitive to say the least.

Because trends are so often portrayed as predictions it is easy to forget that they have little predictive value in a constantly changing world. The demographic transition is the product of a world drastically re-shaped by newly global-scale changes in energy available to human beings and the

35

rate that materials are extracted and burned (consumed) along with other resources like wood and naturally occurring food sources. Many of these changes are almost certainly one-offs.

Yesterday's demographic trends are not predictions and their continuance can be and has been interrupted by many things. After the second world war both France and Australia feared demographic 'implosions'[63] and were taken by surprise by the mid-century baby boom. As recently as the first years of the 21st century, French demographers were again taken by surprise by a return from a relatively low total fertility rate to a higher one.[64]

The demographic transition theory is more an ideological explanation than a true theory, seemingly based on a desire to normalise the demographic curve over the time of the industrial revolution, from about 1750 to perhaps 1973 of the industrial revolution and the demographic experience of the so-called 'first world'. As might be expected, hypotheses so-far based on this have shown little or no predictive capacity.

> "While the theory of the Demographic Transition is of importance to demographers (including those trained as economists) .... For the most part it seems like a grand historical generalization buttressed by a variety of ad hoc causal assertions." *Leibenstein* [65]

> "This overly simplistic and ahistorical view has provided fertile soil for the propagation of demographic transition theory as the appropriate paradigm - implicit or explicit - for the understanding of African demography in the recent past." African Population and Capitalism: Historical Perspectives.[66]

Despite these very serious criticisms, the demographic transition theory is still a major underpinning both of economic growth theory and of global capitalism.[67] And just about everything that humans have ever wanted or needed is being sacrificed to dogma based on these three ideologies.

The most persuasive argument *against* 'progress' is how maintaining the material expansion and population growth it is measured by has become counterproductive to democracy in first world countries, exemplified by the coercive rhetoric of the 'Damn 'em all' article. And, ironically, the most persuasive argument *for* maintaining 'progress' in the first world against all objections is the one that holds that progress brings democracy to the third world.

Another explanation of the Demographic and Economic Growth that accompanied the Industrial Revolution.

The industrial revolution brought into being a feudal empire where the obligations were created through money rather than via blood-relationships and made things even worse for many before it made them better for some. To create the workers and the markets for an industrial economy required dispossession, for no-one with adequate land would leave it to go to sweatshops and imperil their lives and those of their children in dangerous mines in slave conditions. Leaders were placed in positions by war or colonisation where they either lost all or they sold out on their people in order to retain some power and privilege with the new regime.[68]

It was coal and oil that brought wealth so enormous that more people were born than the world had ever seen before and, of those hoards, some were able to lead the lifestyles of rich people at the peak of complex agricultural economies. But those people were always standing on the shoulders of fossil fuel slaves, local slaves or off-shore slaves.

Fossil Fuel Slaves

"Fossil-fuel slaves" is a Buckminster Fuller term of measurement[69] that illustrates how the commercial availability of fossil fuel in the developed world delivered citizens the energy equivalent of many human slaves. An individual who wanted something that required more work than they could do alone but who could not interest others in making a cooperative effort, had to find power from some other source: wood-fire, water, or wind, for instance - usually requiring mechanical technologies. Alternatively they had to capture and force other animals or humans - as serfs and slaves - to do the work they wanted. These cruel alternatives have formed the basis of much of our civilization. The role of slavery in the history of economic progress has been largely papered over.[70]

Access to abundant fossil fuels like coal, petroleum and natural gas eventually meant that people were able to get machines to do the work of many beasts and humans. When a well-resourced suburban citizen from an industrialised country makes toast from shop-bought bread, no coercive effort is required of the toaster. It is no longer necessary for that suburbanite to go out and force unwilling captives, half-starved serfs, or poorly paid servants, to tend the fields, haul the wood, build the fire and make the bread. The fire building, bread-making and toaster slaves have been replaced by electricity derived by burning brown coal. The slaves that tended the wheat fields have been replaced by a couple of people who drive diesel or petrol fuelled moving factories on wheels which do the work of tractors, planters and harvesters. These machines also thresh, winnow and package the grain, which is then taken by truck to silos and put on electric or diesel trains. Later it is distributed by more trucks to central depots; thence, to factories that make bread, which is then distributed to supermarkets in shopping complexes. Our well-resourced suburbanite picked up his daily bread and his toaster in such a shopping complex and took it back to his home in his car. This system is why access to plentiful fossil fuel energy in the 'first world' can be seen as the equivalent of having many slaves.[71]

Parts of the processes of obtaining coal, gas or oil do involve coercion – but it is usually hidden by geographical and social distance or even by historical distance. The industrial revolution had its origins in extreme inhumanity as animals, women, children and men with little or no choice worked in conditions that only just allowed them to survive for shorter than average lives. At the time these conditions were made possible because the sector of society that actually benefited from all this work was able to minimise the perception of suffering and to justify it in various ways. Time has created distance from these conditions in some countries. In affluent countries, with well-established union movements, some miners' conditions improved greatly during the long economic boom from the end of the second world war until the 1970s and 1980s, but since the 1970s oil shocks, dwindling margins and rising costs have seen the erosion of working conditions and old practices of bringing in foreign workers to undercut and displace workers in long-standing mining communities. Corporations evolved through mining and, although in those post-war times of plenty their conduct might have improved for a while even in poor countries, they continue to exploit workers and destroy communities for profit wherever governments allow this, in styles not very different from the 18th century. These hard facts are not something most people think about when they turn on the electric toaster, catch the electric tram, purchase cheap goods from Indonesia, or fill up their cars with petroleum sourced in Nigeria. There is not much they can do about something far down the line in a system in which most of us are completely enmeshed, although 'wirelessly', to a similar degree and without much more understanding than the toaster.

It may be that freedom and democracy can only be widely disseminated in a complex society if there is abundant fossil fuel. Distribution of available wealth is almost as important as its presence since a society where an elite is very rich but a huge mass of poor are cultivated as an adjunct to fossil fuel is neither free nor democratic.[72] Democracy, or at least full participation in activities and decisions, is much more likely in small societies without intensive agriculture.

Per capita oil and per capita economic growth

The fossil-fuel empire that began in the 1750s only really lasted for two and a half centuries. The oil era, the most triumphant period, only prevailed at peak per-capita consumption for the fifty years between the Second World War and the early 1970s, plateauing out from 1979, as the global average birth-rate slowing down coincided with the decline in big new oil discoveries. World-wide per capita economic growth began its jerky and unequal dwindle downwards from 1973.[73] Since then the peoples of the 'third world' have continued to bear the increasing brunt of oil rationalisation, even though their countries have provided much of the first world's oil.

Off-shore slaves

Since much of the world's wealth since 1973 has been mined from distant 'undeveloped' lands,[74] this decline in wealth has been largely invisible to readers and observers in the official developed world. When famine-populations in Africa are paraded on the television, it may seem like these are mysteriously recalcitrant, but presumably temporary, exceptions to the rule of progress. Manufacturing takes place off-shore as well, in countries like China, where 'development' displaces and dispossesses millions, reproducing in magnification the 18th century diasporas of landless workers from rural areas for the factories that make the myth that globalization will reduce the costs of goods for everyone, seem true in the first world. The usual incentives to ignore the injustice and pollution prevail; this is how these people have always lived and 'development' is just about to bring them out of their chronic poverty and despair to replicate our industrial revolution. There can be no turning back –

(Someone has to pay the price of Progress).

If things were to go according to the great Progress Plan, then, after a period of extraordinary waste and suffering, having passed through a communist revolution and an industrial revolution, the Chinese population would emerge, reduced in size (not by the mechanism of the demographic transition but via the one-child policy), increased in individual wealth, and ready to march shoulder to shoulder with the rest of us *developed* peoples towards that human epiphany that lies around the next turn of the great highway of Progress.

What will happen if progress turns out to be a myth built over the reality of finite oil?

What will happen if there is not enough petroleum and coal to fuel the fires of these successive industrial revolutions in China, India, Indonesia, South America, Central America, the Philippines and Africa…?

In *The Final Energy Crisis*, "Battle of the Titans,"[75] ex USSR resident, the late Mark Jones, writes about competition and war between the USA and China over oil, giving the USA only a few years before China is able to mobilize equivalent human and material forces to challenge US world-dominance. In the same book in "The Chinese Car Bomb,"[76] Andrew MacKillop writes of how China's 'car population' is doubling every four years:

"… an insight into exactly why three nuclear-armed powers – China, India and the US – are ever more likely to fight among themselves, or confront EU importers, including two nuclear-weapons states for *the last oil reserves on the planet.* Under any hypothesis – excluding childish technological fantasies and utopias such as those trotted out by Amory Lovins or US Energy Secretary Abraham – there is simply no prospect of China, India – or other countries such as Malaysia, Brazil, Turkey, Iran, Ukraine, Mexico, the Czech Republic, and other emerging car producers – being able to achieve US, West European, Australian or Japanese rates of car ownership. The Chinese Car Bomb therefore ticks onward, as each day another estimated 112,190 cars are produced. Each one requires up to 55 barrels of oil-equivalent to produce, and must operate on bitumen-based highways, on tires that themselves are about 40 per cent oil by weight. Not only is this explosion of the world car fleet a serious threat to the earth's environment, but through its oil demand impact it will become a threat to international peace and stability. In the same book, and X writes about competition and war between the USA and China over oil, giving the USA about five years (of which two have already passed as I write) to slap China down before the US overdraws on its debt and oil reserves and China is able to mobilize its human and material forces to enslave the US. "

MacKillop also invokes the horror of internationally imposed 'Belsen economics' in Africa in "Dark Continent, Black Gold,"[77] where entrepreneurs exaggerate future oil production, gaining foreign investment in their privatized projects, whilst foreign governments ensure through strings attached to foreign aid that Africans themselves never benefit from their own reserves.

And running out of petroleum and even coal (as we draw on our reserves to replace oil) is not our only problem because, ironically, we are likely to fry or freeze, due to global climate change, and choke to death at the same time, from a combination of US emissions and those that will come from the Chinese Car Bomb. So, even if petroleum had lasted forever, as the demographic transition supposes, civilization would not survive its fumes.

That leaves some alternative fuel that doesn't pollute and which will be available in the same ginormous quantities as fossil fuels. Nothing fits the bill as yet although we are promised that clean coal, hydrogen, breeder reactors, nuclear fusion will.

The Future of Progress in two different world systems

So, what does this mean in the scheme of the things that began with the British Industrial Revolution?

Well it means that until and if we find some alternative to fossil fuels that the populations and activities of all countries will need to come right down to below where they were when the great British Industrial Revolution began. And that means that the world will shed something like 5-8 billion people if no new magic solution comes along. But the United States and its satellites – Australia, Canada, New Zealand – and England – seem to have growth built into their systems and no brakes. India, Saudi Arabia, Israel, to give some examples of still 'developing' countries with burgeoning populations also have no brakes.

Some areas of the world do seem to have realized that we cannot rely on 'progress'. Since the first oil shock of 1973, with the curious exception of Portugal,[78] the non-English speaking countries of Western continental Europe and Japan, have sharply reduced their population growth rates. Portugal has a different inheritance system which results both in population growth and strong inequalities. It is in those areas of the world that have reduced their population growth rate that hope for the rest of us may lie, if we can identify what it is about these countries that permits them to recognize, predict and respond to changing circumstances. That is one of the aims of this book and series.

The Future of magic

The people who invent and publish explanations about Progress which motivate 'leaders' to drive the rest of us ever further down the one-way road of good intentions have forgotten - if they ever knew - that our industrial revolution relied on coal and oil. If you analyse the logic behind the myths of Progress and economic growth, it seems that most people actually believe that the industrial revolution was entirely a product of human ingenuity. We can infer this because the same people object to concerns about finite resources by saying that human ingenuity will always find a way. This is tantamount to saying that the technology and vast wealth of the industrial revolution were due to magic. It totally disregards the role of fossil fuel reserves, which formed completely independently of human ingenuity but without which 'modern society' could not exist. It is as if religious myths of humans reaching perfection and lightness of being have been confused with earthly reality and that humans have implicitly assumed that they are becoming celestial beings with magic powers due to their unique qualities as a smart or divine species. For the unscientific, modern civilisation must be like an Indian rope trick, or reliant on moral and spiritual bootstraps. Trains, rockets, cars, factories, electric lights are powered by magic. Willing cows, crops, and plastic trees fill the supermarket shelves with milk in cartons, bags of flour, and implements and containers made of plastic. Petroleum mysteriously replenishes petrol pumps on demand. The shops are full of sparkling trinkets like Aladdin's caves. You can pick up a plastic implement and magically talk to someone thousands of kilometers away whom you have never seen. You can press a switch and a virtual world of little laughing, talking, singing, important people appears on a glass screen. You can make all these things happen at once on your computer and on a hand-held device. The quality, quantity and rate of delivery of the magic goodies – particularly the electronic ones from the industrializing countries - seems to be intensifying, but per capita oil has apparently been on slightly variable annual plateau since 1979, if we are to believe the published data. Closer examination reveals that the definition of 'oil' has changed and that our fuel sources are increasingly leveraged and socially rationed by price. If we do not personally feel the impact of cheap petroleum energy decline per capita, it is probably because we are in a socially buffered class.

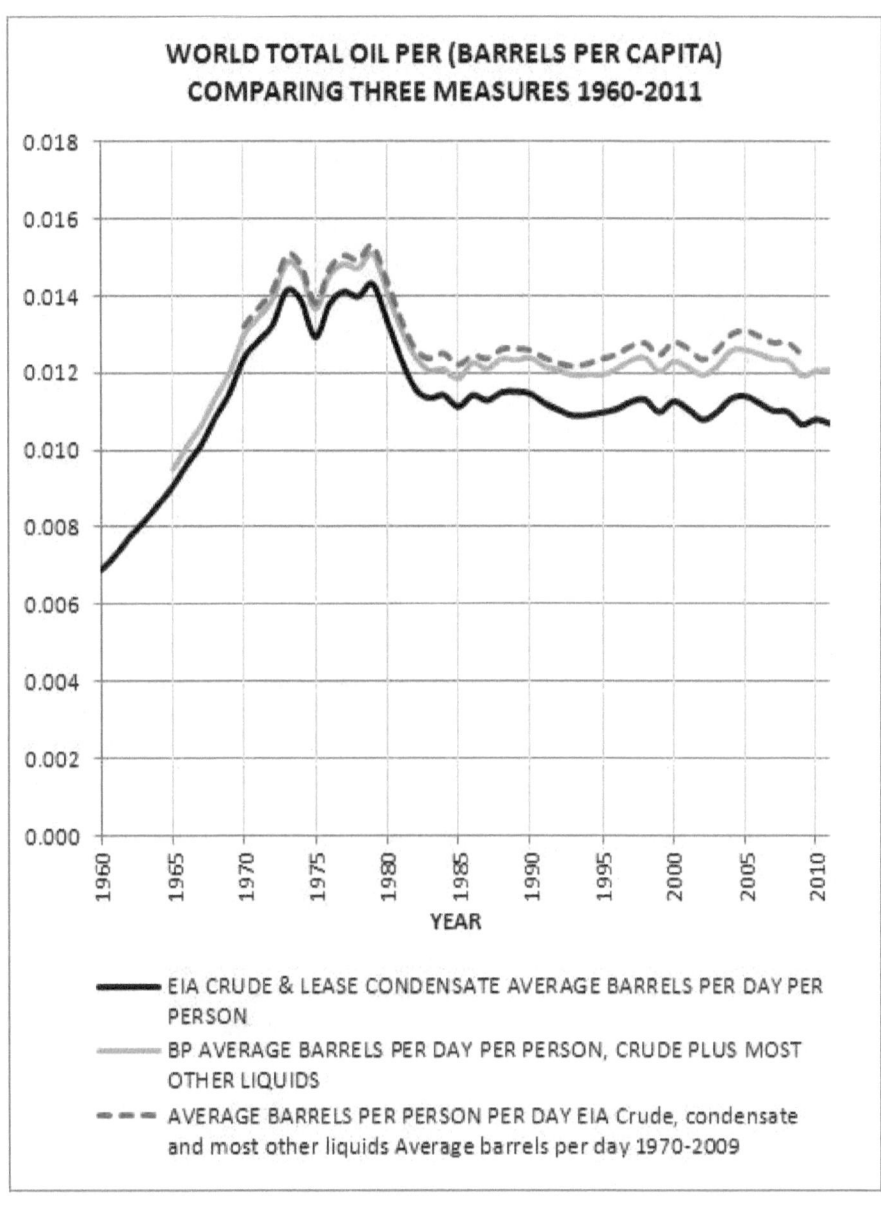

Figure 3. World total oil per capita 1960-2011

Figure 3, 'World total oil per capita 1960-2011' suggests that, where once oil came from oil wells, now, to keep up with human population growth we supplement well-oil with all kinds of other stuff, whilst still presenting it as the same old stuff. If we still relied just on crude and lease condensates from oil fields, where trends are shown by the lowest and oldest line in the graph, we would be considerably poorer in total oil supply than we seem to be. Early definitions for total world oil (see EIA Crude and lease condensate from 1960-2011) only counted crude oil and lease condensate. Coinciding with the decline in easy oil availability as it became necessary to look harder and deeper for oil, the definition of oil started to include other sources obtained away from petroleum fields. The BP definition above counts crude oil, shale oil, oil sands and NGLs (the liquid content of natural gas where this is recovered separately). The EIA definition for Crude, condensate and most other liquids from 1970-2009 comprises natural gas plant liquids, and 'other liquids', defined as, "Biodiesel, ethanol, liquids produced from coal and oil shale, non-oil inputs to methyl tertiary butyl ether (MTBE), Orimulsion, and other hydrocarbons." That's a lot of new sources, all of them requiring more energy to extract than that required to harness the traditional 'gusher' close to the surface, now a rare phenomenon.

Highlighting the increasing cooption of non-crude oils and other processed and biological materials to 'total oil' makes the decline in crude and lease condensate more obvious, but the strange wavy plateau in the graph still gives the impression that we remain in control of our destinies. Some economists even believe[79] that we are still getting richer by producing more for less, that economic growth has become 'dematerialised' from fuels due to scientific efficiency. There was some truth to this shortly after the oil crash, but those gains are not as great as they are touted.

Another explanation is that, as prices for oil and other fossil fuels have risen, our efficiencies lie more in carefully choosing the kinds of fuels we use for different purposes, keeping oil only for the tasks that cannot be done well without it. For instance, where people once used oil to heat their homes, they now use coal-fired electricity or gas.[80]

We are told by politicians that flow energies like hydro, wave, solar and wind are contributing more and more to energy supply, but their contribution is small so far and can never match the vast reserves of fossil fuel that modern civilization is based on, because they cannot be stored in significant reserves and they tend to change on a daily basis and to be unpredictable.

The major alternative sources that are really supplementing crude production are mostly pretty nasty, with a major symptom being clashes over democracy, even in the 'developed' world now. Farmers and communities are losing battles against corporations and their own governments over gas-fracking that threatens water supplies.[81] Mountaintops are being blown up to extract coal.[82] Tar-sand and shale-oil mining are last resorts that devastate land, use more fuel to extract, give less energy, and pollute atmosphere and water more than conventional sources. The world's diminishing forests and their rare animal inhabitants are bulldozed to grow soybeans and other food crops to replace conventional crude and gas. Not really signs of increased 'efficiency'.

Maybe this is really a bootstrap operation after all.

Although nuclear energy is important in electricity production in many countries, those countries got into nuclear at a time when there was plenty of capital. With some exceptions, like Iran (which encounters international political resistance) countries that do not already have it will have difficulty financing such massive new operations in global economies that keep faltering along with our crude oil supply.[83]

According to Figure 3, even with the addition of oil from these new, expensive, pollutive and environmentally destructive sources, total global production of 'oil' is only just keeping up with population growth, if we take .012 barrels or so as the norm since 1982. Except – everyone doesn't get their .012 barrels.

The cost of oil is increasingly unaffordable.

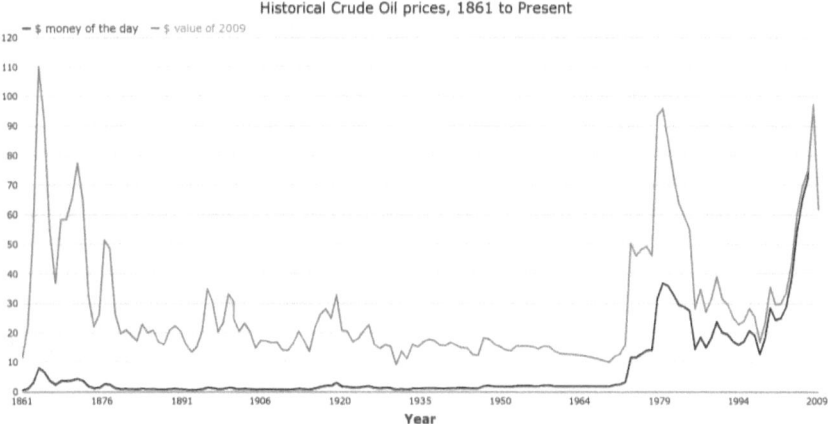

Figure 4. Historical Crude Oil prices, 1861 to 2009
Source: http://chartsbin.com/view/oau Average oil production per person does not reflect the reduction in oil availability in poor countries where people are not able to pay increasingly high prices for oil, even for

cooking fuel needs, leading to riots, especially in 2008.[84] **We do know that more people have less oil and fewer people have more oil.** [85]

What does seem certain is that the rich have NEVER been richer.

In the developed world, the gap between the rich and the poor in the first world is increasing, especially in the English speaking countries.

In Africa, in South America, in the islands of the Pacific, including New Guinea and the Philippines, whole peoples continue to lose their homes and their lives so that an ever diminishing proportion of the human race may continue to believe in magic.

But to assume that all Americans, New Zealanders, English and Australians are wealthy citizens of 'wealthy countries' does many a disservice. These countries are increasingly socially segmented between the land-rich and the land-poor, with the high cost of rent eroding living standards and small to medium business profits. Richard Wilkinson and Kate Pickard in *The Spirit Level* situate the beginning of the growing divide around 1980 in their study of wealth distribution within 26 different countries.[86]

> "Within countries like Japan and some of the Scandinavian countries ... the richest 20 per cent of people are less than four times as rich as the poorest 20 per cent of people. At the bottom of the [scale] are countries in which these differences are at least twice as big, including two in which the richest 20 per cent get about nine times as much as the poorest. Among the most unequal are Singapore, USA, Portugal, the United Kingdom, Australia and New Zealand."[87]

Wilkinson and Pickard make no attempt to explain why Portugal should stand out in such a contrast to its continental European neighbours, which almost invariably outperform English speaking countries on equality and social conditions. I would suggest however that one reason lies in Portugal's inheritance laws, which reflect a tendency to disinherit children in favour of spouses, similar to English-style laws. In the other European countries Napoleonic-style civil codes provide for children before spouses equally, therefore men and woman come to adulthood with similar wealth. The spouse-preferring inheritance (and divorce-settlement trends) of the Anglophone countries and Portugal are recent attempts to compensate systemic sex-bias in property ownership, but they do not help much in the overall tendency for property to accumulate in private hands in these systems, which is a major study of this book.

All the signs are there in new land-enclosures – the selling off of public land to private corporations, foreign persons, and local nabobs; the speculation that goes on among the propertied class, excluding an increasing number from housing and from choosing self-sufficiency on their own land, condemning others to the precarious situation of working for a class of difficult and stingy employers whilst paying that same class as much rent as it can extract.

The Progress mythology accompanies these changes in what Veblen[88] called the "robber culture." The mainstream Anglophone media has latched onto the Calvinist American Dream and teaches people to believe that everyone has a good chance of making it to the top and, if they seem to be slipping downwards, it must be their own fault. This is no different from the messages that have emanated from pulpits for centuries. Unhappiness, failure, and poverty are sins. Those who suffer from these shameful attributes naturally try to hide them.

In all this the real magic – by which trees and other plants make energy out of sunlight and soil – has gone off the official political human radar. Economists once explained how society was built from what, beyond mere survival, humans could squeeze out of nature. Now economists act as if they believe that humans can force nature to produce whatever humans need and more and more pressure is placed on people to come up with answers to problems that are not even clearly stated. We are cajoled and forced to give up rights, wages, conditions and sovereignty for an illusion. The illusion is that things will go back to the way they were when fossil fuel was plentiful and humans were not so multitudinous, but no official authority is given to the reality that fossil fuel is not being replenished and we are running down supply at an accelerating rate.

Let us be wary of the doctrine of Progress.

Need I add that we should be wary of the doctrine of progress? The 8 hour day was won in the 1860s and the last time that a human walked on the moon was in 1972. We are losing our hard-won industrial rights, the era of the car is receding, and coal is the fuel of the future again.

In fact, as we stated in our introduction, two different world systems exist, with different outcomes for population numbers, energy use, land distribution and democracy. Before we look at them, however, we will explore some population theories.

CHAPTER 2: MODERN FERTILITY, MORTALITY AND DEVELOPMENT MYTHS

Factors blocking change

Researchers in the expansive Anglophone economies are locked into the present system by a vast infrastructure of journals, conferences, books and mass media, wholly or partially devoted to funding the spread of industrial capitalism. Societies of volunteers and professionals are organized around the belief in chronological human progress. Projects are marketed specifically for development, and funding earmarked for research into this area. To reject the validity of progress is to reject these sources of position, status and money. Progress is at the very heart of capitalism and its antithesis, communism, because it is the flagship of modern economics. If economists turned out to have been wrong about progress then what else might they be wrong about?

A classic example is provided in John C. Caldwell and Bruce K. Caldwell, "Pretransitional Population Control and Equilibrium."[89]

> "A persistent theme in much anthropological writing is the concept of the deliberate control of population numbers by hunter-gatherers as a means of achieving moderate family size, adequate nutrition, and constrained adult mortality. An analysis of the mix of theory and field evidence that led to this conclusion finds the case not proven. On the contrary, Malthusian constraints can operate, and probably did operate, to produce a hunter-gatherer society where most adults were reasonably robust and healthy even though child mortality was high and life expectancy short. The absence of population limitation in pre-Neolithic times implies high mortality as well as high fertility, and weakens the argument positing a Neolithic mortality crisis."[90]

In this article, the authors write as if they believe that hunter-gatherers lived for only 35 years or so. Because they see no consistent evidence of mechanical contraception, but acknowledge stability in hunter-gather numbers, they assume that there was high mortality through violent lifestyles and environments. The authors argue that anthropologists have based their arguments on unreliable and occasional data about the availability of contraception in simple societies and have extrapolated beyond this. They point out a data definition difference, where anthropologists may have counted infanticide as contraception but demographers would count it as mortality. The authors do not acknowledge the almost universal occurrence of social practices that reduce fertility opportunities. These are well-documented by anthropologists and are an important theme of this book, which will ultimately also identify other factors – notably incest avoidance and the Westermarck Effect - not previously considered from this angle, for their impact on population growth rates.

Why are there such glaring contradictions between observations made by anthropologists during first and early contacts with Pacific Islanders and assumptions made by late 20th C missionaries, family planning and foreign aid commentators and theorists? The early work was outstanding. Much of the new work is dross.[91] At issue is the duration of Pacific Islander societies, methods of family planning and population control, life expectancy, and other demographic norms.

'Modern' theory seems to be history-poor, based on assumptions that make little sense. Late 20th century practitioners of aid based on the *demographic transition* seem to be the most blinkered. Without some other explanation, this demographic transition theory appears to serve industrialism, globalism, religion and professional NGO officers far more than it serves the peoples of so-called developing countries, or democracy in the developed countries, where immigration from poor countries is used as a source of cheap labour. But to practice development-aid requires an 'undeveloped' target; the industry would disappear if it were conceded that the targets' situation was usually much better before colonisation and 'development'.

There is abundant evidence and below I include excerpts from Peter Pirie,[92] "Untangling the Myths and Realities of Fertility and Mortality in the Pacific Islands", to illustrate that massive amounts of crucial data have been discarded and ignored in the field of anthropology alone. Furthermore, this data has been replaced with what amounts to racist colonial propaganda, used by people whose principles would run completely counter to such papering over realities of suffering, genocide, and dispossession.

Pirie uses early anthropological studies to derive information about typical 'pre-transitional' population attributes and finds that steady state societies were common. He gives the example that only 28% of the Kunimaipa people in the Papua New Guinea highlands in the 1950s were under fifteen years old. There was a child/woman ratio of only 380 per thousand and, at the end of their childbearing life, Kunimaipa women produced 3.5 children on average. This rate of reproduction, given the mortality rate which was high for children at 225 per one thousand births, apparently kept the population in balance and at steady numbers. The Kunimaipa people used typical strategies to reduce family sizes, including abortion, post-partum taboos, breastfeeding for four or five years, and infanticide.[93]

" Margaret McArthur (1961:7-12) noted that, while abortion, infanticide and infant mortality were not deliberately concealed, mothers responding to fertility questions usually omitted these events as the outcomes of pregnancies that never became "people". Although the numbers on which these indices are based are too small in this case to be statistically reliable, other studies have shown similar results and there can be no doubt of the biological situation and social reality they describe." [94]

Ron Crocombe's important 1971 book, *Land Tenure in the Pacific*, contains seminal contributions showing well-known and lesser known islands before and after colonisation. These case studies include Australia, where, in 1953 Joseph Birdsell produced a quantitative ecological analysis which demonstrated the importance of a single environmental variable, mean annual rainfall, in determining the size and population density of tribal areas in Australia. He used basic data recorded by Tindale in 1940, who had mapped the territorial boundaries of about 400 dialectical 'tribes'. Birdsell wrote,

"A single environment determinant cannot completely express the relationship of the numbers of men to their land, but in this case a coefficient of curvilinear correlation of 0.18 was obtained. This close relationship was determined for basic ecological units, the 123 dialectical tribes whose resources primarily depended upon locally earned rainfall. The mathematical function of the relationship was logarithmic and areas decreased and densities increased as the mean annual rainfall rose. The analysis showed continuous variation, as opposed to stepped variation, between the minimum rainfall of four inches, through the total range to the maximum of 160 inches annually in the rain forests of north-east Queensland." [95]

Yet the idea persists that traditional Pacific Islanders spent 40,000 years or more in a classic Malthusian struggle, due to ridiculously high birth rates. Pirie writes on "the myth of traditional fertility", whereby, after near die-offs, populations in colonised islands started to blow out:

"This experience led to the identification of another myth: that these recently exposed fertility levels were "traditional". The large family size noted in so many populations in the Pacific (see table 1) was thought to mark a return to normal levels, due to the success of departments of health in reducing diseases that had previously been so debilitating. Yet average family size approached seven children, with a third of all women with completed families reporting 10 or more children (up to 24 in the aforementioned case)." [96]

Pyrie draws our attention to the obvious which is unaccountably invisible to development and foreign aid commentators, that is, that

"Such fertility could not have been sustainable in a population in residence for something like 3,000 years, living fairly peacefully on small islands largely free of environmental and contagious disease. Some other influences [that is, fertility constraints] as yet undiscovered must have been at work [in former times], but certainly were not operating in Samoa at that time". [97]

He remarks how the norm had become for men to boast about the numbers of children they had and for women to be judged for their breeding capacity. This was, however, a reflection of industrial society norms brought by the colonisers.

"Victorian colonialism, the churches and the Samoan way of life (fa'a Samoa) had combined to set pronatalist standards for the population." [98]

In Britain from the middle of the middle-ages and increasingly during the industrial revolution and the era of Victorian colonialism, dispossession combined with poor laws to boost birth-rates in the labouring and working classes. See Book Two of *Demography, Territory and Law* on Land Tenure and the Origins of Capitalism for further explanations.

Early colonialism meant an initial die-off for many islanders. Colonisation was a terrifying time when land that had for centuries been apportioned to provide each islander with sufficient, was arbitrarily reassigned for new economic purposes over which the islanders had little or no control. Of course this meant poor nutrition, starvation, violence (as the colonisers seized and the islanders attempted to defend what was theirs) and marginalisation. If the population survived at all, there was often a rebound. Certainly the islanders were encouraged to have children because the colonial economies needed unskilled labour. Few island economies can sustain economic growth for long though. For a while the economy coped with large families, soaking up workers. The people who had survived and then multiplied enjoyed the novelty of consumer goods. Although their health, status and economic position deteriorated, some health care, schooling and housing were initially provided in paternalistic colonial societies.

And then suddenly there was a 'population problem'. The Pacific Islands and their peoples are now stigmatised by their terrible population problems.[99] I sometimes comment on the unfairness of this stigmatisation to Pacific Islanders, such as indiginous people from New Zealand. I have also been interviewed on the subject by an Australian Aboriginal radio program on Special Broadcast Service (SBS) Australia. The impression I am given is that Pacific Islanders of direct descent deal with difficulty – with a kind of disbelief in both - with two conflicting views of history. One is that their peoples were hopelessly disorganised, incompetent bumblers until colonisation came to rescue them, and the other is that their ancestors, to the contrary, must have had it all sewn up because, as Pirie puts it:

> "… it has been observed repeatedly that densities in the Pacific islands never seemed to rise to the levels at which their "carrying capacities" were strained. Resources were appraised very conservatively relative to the population dependent on them and their rate of utilization; subsistence agriculture, with few exceptions, remained casual or non-intensive."[100]

Islanders may find this perspective on their traditional demography refreshing and validating, but such observations strain the credulity of most Europeans, who simply have not got the knowledge of history and different systems to question how the present situation came to be.

This strange naivety also dominates the paradigms of many contemporary professional writers on Pacific Islander issues. Pirie writes of this bizarre lack of critical appraisal or knowledge of history in recent learned literature. He is writing about articles published in the early 21st century.

"The obvious alternative explanation was that fertility must have been controlled. Given the prevailing ethos [in academic literature], this possibility seemed too far-fetched. However, over time, and with more reading about other cultures in the Pacific and the world, this explanation became more convincing to the author. In summary, the control of fertility involved the imposition of a variety of traditional methods to extend birth intervals and limit family size"[101]

But he has to go back a long way to find data. Yet there is a lot of this data. How did it all get lost? How did it disappear for all practical purposes from the radar of a legion of organisations, funding bodies, experts and academics who sojourn, study and man foreign-aid and family planning organisations in the Pacific? The answer, I fear, is that it was buried under the voluminous publications testifying to chronological 'Progress'.

At any rate, as Pirie writes,

"One of the first cases to appear in the literature was described by Firth (1936) who found such a situation on Tikopia, a Polynesian outlier in the eastern Solomon Islands. Because of limited space and resources, the need to control family size is likely to appear most obviously on such an island. Two other cases have been noted on atolls, Nukuoro and Eauripik, in the Federated States of Micronesia. Both have retrospective data of good quality that show low mortality, an expectation of life of about 60 years, and apparently low fertility to have resulted in population homeostasis over a long period up to the first decades of the twentieth century (Carroll, 1975; Levin, 1976). However, control of family size was not confined to areas of such limited possibilities for subsistence, but was prevalent - sometimes in rigorous form -in areas as diverse and as environmentally productive as the highlands of New Guinea and Tahiti."[102]

Pirie goes to some lengths to justify his observations[103] which should not require such effort, since the historical evidence is so abundant. But Pirie is writing to overcome a psychological dead weight of prejudice. One feels that he is shouting loudly but that his voice is muffled almost into silence by the colossal inertia of prejudice that is the Anglophone world's colonial heritage. And when I use the word 'colonial' I am talking not just about our attitude towards Pacific Islanders of non-European origins, I am talking about our understanding of our own origins. Two crucial truths need emphasis here. The first is that our pre-medieval European ancestors – Celts, Iberians etc. – lived in steady state societies by the same rules that kept most Pacific Islander societies comfortably within limits. The second is that many of our own ancestors were exploited in just the way that the Pacific Islanders were. They were dispossessed and then encouraged to have large families to provide labour for the convenience of those who dominated increasingly complex societies. Our understanding of history has been coloured by the same pragmatic explanations to justify exploitation that have been used to confuse today's Pacific Islander.

As well as the fertility myths, there are, as Pirie describes them, the 'mortality myths' – widely accepted ideas that the life-expectancies of pre-industrial peoples were indicative of extremely short life-spans - which have been used again and again to justify hardships associated with our own industrialisation and that of other peoples. Without these justifications much of what is done in the name of modern development, in the first world and the third world, looks at best like bare exploitation, at worst, like frank land-stealing and pillaging.

Contrary to the myth of the degraded, unhealthy, scrawny savage, the first European explorers found the South Pacific peoples to be relatively free of disease. Pirie notes that the explorers brought with them "bubonic plague, smallpox, tuberculosis, cholera, dysentery, yellow fever, leprosy, syphilis, influenza, measles, whooping cough, diphtheria, scarlet fever, typhus, typhoid fever and gonorrhoea, to name some of the more destructive diseases. Several diseases specific to tropical areas, such as trypanosomiasis, or "sleeping sickness", also were absent..." although malaria, he adds, was present in Western Melanesia as were some other mosquito borne diseases, such as elephantiasis. Yaws was also a problem.[104]

There was however little evidence of premature death, by accident or by warfare. The climate and environment of the inhabited Pacific Islands is, as Pirie observes, "among the world's more benign."[105] Pirie reminds us that traditional warfare was "a rather protocol-laden affair in which causing massive fatalities among the enemy was not the main objective."[106] Where indigenous peoples might sometimes seem to industrialised Europeans as overly emphatic in their gestures when drawing boundaries on behaviour, we should remember that, in the absence of the specialisation which, in our own societies, has resulted in police and armies, locks and keys, in a small non-complex society, the individual has to deal with aggression personally. They need to make their intentions perfectly clear before they get too close, for there are no hospitals and ambulances.

Pirie draws the obvious conclusion.

> "Could it be that most Pacific islanders finally died on their mats of heart attack, cancer, or stroke?"[107]

He observes that,

> "Demographers find such a possibility hard to accept, for it is at odds with the theory of demographic transition, which assumes that the collective lot of a pre-transitional population was usually a swift, nasty and early end. Norma McArthur (1977:273), for one, wrote that that possibility "strains credibility" when this author first put it forward, but no one has been able to come up with a more convincing alternative."[108]

Pirie is right and most of the others are so wrong. And so much harm has been done and continues to be done to all of us by these myths of fertility and mortality.

But Pyrie, in noting indigenous methods of population control, in his paper, does miss out on the remarkably constant quality of incest avoidance and the Westermarck Effect in the fertility regulation of Pacific Islander and other human societies and in the population spacing of other species. I will explain more about this function of incest avoidance and the Westermarck Effect after the next section where I review other critical theories about human population dynamics.

Evolutionary theory of human life span – Kaplan et al

Kaplan, Gurven and Winkling's (2009) Evolutionary theory of Human life span[109] seems to go beyond most of the limitations of other recent biological ecological or evolutionary theories.

On the subject of how long humans lived, Kaplan et al argue that humans had potential to live for about 70 years in hunter-gatherer environments.[110] They theorise that reaching age 65 must have played a part in human evolutionary adaptation for the acquisition of life learning and storage of environmental knowledge over long periods of time and distance in elderly humans.[111]

On the subject of causes of mortality, they assert that the comparatively isolated populations that persisted in localized environments pre-agriculture and large-scale settlement would have become genetically adapted to local pathogens. The effect of this would have kept disease mortality low.

They point to studies which strongly suggest that in previously isolated populations local immunity probably existed to small pox, falciparal malaria, tuberculosis, cytomegalic virus, Epstein-Barr virus, pneumonias, herpes, hepatitis B and arboviruses. An implication here is that population movements across space would have caused exposure of naïve populations to such pathogens with new epidemiological consequences including epidemics and chronic disease.

In *Virolution*,[112] Frank Ryan hypothesizes and gives examples of evolutionary cooperation at a cellular level between local biota (including humans) and local viruses, where humans and other lifeforms don't develop diseases in the presence of viruses that cause disease in other populations of the same species. Exposure to these viruses within the same species but involving populations from different localities typically gives rise to illness and mortality. This kind of effect is also known as 'virgin soil' epidemics in anthropology, where for instance, Spanish invaders brought smallpox to South American indigenous peoples, but succumbed themselves to local tropical diseases.

Ryan's theory is helpful to understand why diseases manifest later in life as the autoimmune system deteriorates. Human examples may include shingles and rheumatoid arthritis among many familiar diseases where risk and severity increase with age.

Kaplan et al's theory on mortality in well-established isolated populations and Ryan's theory run very much counter to the popular idea that modern medicine caused a drop in mortality, since such theory does not take into account a likely increase in mortality that resulted from massive population movements, such as invasion, immigration, colonization, land clearing and rural to city drift.

Kaplan et al also theorise genetic interaction with environment to regulate diet, work, energy budget and reproductive behaviour.

> "Genes (…) that control defenses against pathogens, repair of cell damage, and reproduction interact with those environmental assaults to determine population distributions of individuals of different ages and their associated physical states at the cellular and organ levels. Genes also interact with environmental conditions (including distributions of energy in the environment and production technologies and with physical state to determine diet, work, energy budget, and reproductive behavior. Those behavioral patterns have feedback effects on physical state since work exposes people to risks of injury and physical stress but also provides energy to support repair, immune defenses, and reproduction. Both changing physical states with age and behavior result in mortality and reproductive schedules with age." [113]

Kaplan et al are talking about the evolution of the human race and other organisms being regulated by genes interacting with the environment over the long-term, in a 'Human Adaptive Complex,' associating with our species a particular life span and allocation of energy across age groups due to a variety of feedbacks they hypothesise. For instance they hypothesise that humans differentiated from chimpanzees in a high consumption of meat which they estimate as between 66% to 75% and that this consumption of rich food sources coincided with the anatomy of our brains and the development of human intelligence. They also observe that hunting requires a very high degree of skill which does not reach its maximum in human beings until they are in their mid 30s.[114] In short they are talking about the shaping of our species over long periods of time through genetic responses to environmental input and how this state is maintained through algorithms (a sort of genetic instruction manual of anatomical, physiological and behavioral responses to many situations and their variations) particular to our species.

Short Review of Theories about Human Population Dynamics

Since I began writing this book, some new work has appeared, endorsing the idea that hunter-gatherers tended to have stable populations and seeking explanations for the coincidence of the rise of agriculture after the shrinking of the glaciers in Europe 10-12,000 years ago and the rise of human population numbers globally.[115]

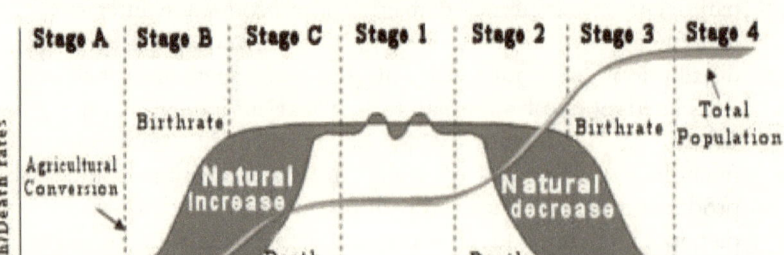

Hopfenberg, R., Hopfenberg, E., and Salmony, S., 2011,
"The Expansion of the Classic Demographic Transition Model"
http://www.panearth.org/OPSO%202011/player.html

Figure 5. Panearth Comprehensive Demographic Transition Model

One example is Hopfenberg and Salmony's "Expansion of the Classic Demographic Transition Model."[116] This model aims to reexpress the 'classic' demographic transition model with four stages instead of three. The classic model begins with rapid population growth exceeding death rates, reaches a stage where death rates rise to meet birth rates, followed by a period when birth rates decline and so do death rates, until, in the third period, long lives and low birth rates are supposedly achieved. Hopfenberg and Salmony's expanded model starts with hunter-gatherer society and low death rates and low fertility rates. It is not clear in either model why the last stage should be stable. The explanation relies on the idea that humans have the power to decide their own family sizes and that they stop having a lot of children when it costs too much to have them. The model does not take into account high immigration for economic motives, through war, colonisation or natural disasters that might prevent stage 4 occurring, despite the wishes of the incumbent population, although it does suggest that hunter-gatherer stability may be overwhelmed by encroachment by other unstable populations. This kind of model is very broad-brush. It does not consider variations in different populations; it attempts to discern a global model. Such a model only makes sense in human populations that respond to global stimuli with similar reactions. It does not allow for populations at a sub-global level that behave differently from the model and from each other.

Human inventiveness

A slightly different example is Craig Dilworth's *Too Smart for our own Good*, which argues that agriculture and an innate tendency to innovate to avoid the immediate consequences of population and economic expansion, eroded our tendency towards steady state societies.

Dilworth acknowledges that hunter-gatherers could have stable societies, but sees overall population growth from Neolithic times onwards, and expresses little hope that stage 4 will ever occur, except in the form of total societal collapse as resources run out and pollution overtakes us completely in a culmination of entropy. He refers to a number of sources exploring the probability that hunter-gatherer populations may have overshot in times past by over-predating windfall abundance in animal populations, leading to sudden population crashes in their food sources and themselves. Between 40,000 and 10,000 BP there was probably a lot of population growth among Neolithic peoples due to their expansion and overexploitation of large mammals, which Tim Flannery called 'future eating'.[117] When this food source became rarer, agriculture may have become unavoidable for some.[118] Dilworth's primary theory is that human beings always come up with some clever technology or practice to overcome new scarcities, but he does not expect overpopulation to survive global resource depletion.

Dilworth's theory assumes that the human sex drive has a function of maximizing numbers of humans and, implicitly, that humans (and all in all societies) tend to have strong sex-drives.[119] The problem then is to explain how, in the absence of modern contraception, the numbers in hunter-gatherer societies remained steady. Dilworth relies for his explanation on male mortality related to territorial conflict, including cannibalism,[120] but particularly on frequent practice of infanticide and abortion.

> "Birdsell for example estimates that between 15 and 50 per cent of all live births were eliminated by systematic infanticide during the Pleistocene as part of the cultural machinery for creating equilibrium between population and resources. (...) Fekri Hassan has suggested that an abortion or infanticide rate on the order of 25-35 per cent would have been necessary to account for the difference between potential and real growth rates during the Pleistocene. And Divale has suggested that a whole complex of population control mechanisms, including systematic female infanticide and feuding, may have been universal to human populations for at least the last 70,000 years."[121]

Although these birth and death control features certainly have their place, they are not the whole story. Dilworth gives little thought to the impact of gender divisions[122] and no thought at all to the impact of incest avoidance or the Westermarck Effect in reducing fertility opportunities, although these are dynamic explanations that I will later pursue, for the phenomenon of territoriality as a limiting factor, which is important to his thesis.[123]

Although he notes the presence of some internal checks (such as aging) to population growth, Dilworth thinks that the main checks are external. Dilworth perceives that humans are behaving like r-selected species, maximizing numbers of offspring over their survival rather than having a smaller number of offspring but carefully nurturing each to increase their survival. Human societies have continually developed new technology that overcomes these checks. If we keep this up we will overshoot all resources and then risk extinction.[124]

Dilworth does not acknowledge that not all human populations are behaving in an r-selection way; his approach is global. His explanations above do not acknowledge that some agricultural societies as well as hunter-gatherer ones were able to maintain population stability, and so they fail to explain how they might have achieved this. None provide an interspecies theory, nor do they take into account the impact of different transport technologies in providing more fertility opportunities.

Pyrie and the other authors – even Malthus[125] – whom I will mention in my chapter about Incest Avoidance and the Westermarck Effect and the Pacific Islander rule, show that there have been steady state agricultural societies as well as hunter-gatherer ones. Although I agree that the rise of agriculture accompanied rapidly increasing population growth 10,000 years ago, I will be arguing in Book 2 of this series that climatic factors and land-loss may have caused an increase in human fertility opportunities, resulting in population pressures, and a widespread resort to agriculture to offset this situation.[126]

Economic Demography

Economic demography writers will usually acknowledge vaguely that natural fertility control processes must have existed prior to the agricultural revolution. An assumption follows that these processes have since been obscured, discontinued, or were never very effective. In this paradigm, a human population explosion began around the time that humans began using agriculture and another, much bigger explosion, began around the time of the industrial revolution. Theories about the reasons for the associated overpopulation and/or poverty began circulating even before Thomas Malthus.

Modern economic and political theory is based on the idea of natural laws of the past from which humans departed by forming a 'social contract'. Hobbes is famous for asserting that life without government, in a condition which he called the 'state of nature', would mean lives that were nasty, brutish and short.[127] He thus justified a hierarchical order as preventing people remaining in continuous ferocious competition with each other.

 Other philosophers believed that these natural laws comported a peaceful stability. Anarchists like Bakunin and Prudhon wanted to return to old rhythms and scale of local communal society. Marx and Engels, reacting to the poverty of masses in industrial capitalism, hoped to find some new stability through communism,[128] but on a similar scale to capitalism.

Locke derived his (capitalist) theory from his interpretation of natural laws, understood modern government as the outcome of conflicts arising from population pressures, but saw these as inevitable.[129]

Underlying all these debates was competition for land. As land-use intensified and population grew, capital became available to exploit land that was once quietly inherited and used simply for living.

Artificial energy accumulation systems as breaks from earlier demographic patterns

Agriculture and industrial society were breaks in a natural pattern of hunter-gathering. Economics and demography have frequently attributed these changes to human ingenuity alone. The role of topsoil or fossil fuel to provide for the large, energy-intensive complex systems we have learned to build has been overlooked. The assumption is that these changes are evolutionary in our species, amounting to a belief that our brains and motor-skills have been constantly improving to produce technologically sophisticated societies. From this has followed another belief, that the natural intelligence of humans as a species will always locate new sources of energy when the old ones run out.

In industrial societies, most humans have gone from actively hunting, gathering and growing food for the body, to getting much of their energy indirectly via a human built environment. The built environment redirects energy from the natural environment via an artificial restructuring. This artificial environment includes the reorganisation of natural groupings of other animals and plants.[130] Such structures as houses, roads, dams, stores, electricity grids, farms etc., feed the human body and supply all or many of humans' other needs indirectly via the energy intensive structures of the society that the body is a part of.[131]

Social reorganization of land creates labouring classes

This energy intensive infrastructure gives the body an artificial environment. But it also breaks up the human relationship with the land in ways that give some humans more power than others over their environment and over each other than they would if the distribution of land and other natural resources had remained direct as it has for other creatures of our size and natural social organisation. The social reorganisation which has accompanied the intensification of energy-use was not the choice of all those who participated. As noted above, this situation has been commented at length by Marx and Engels[132] (Communism), Bakunin[133] (Anarchism), who criticized it, and Locke[134] (Capitalism), who defended it. The origins of the working class lie in dispossessed peoples - captured slaves and infants adapted for servant purposes. (See Book 2 of this series for a detailed exploration of theory and history.)

Durkheim observed the tendency for human society to become increasingly segmented with different segments specialising in roles which all members of simple or flat tribal societies used to fulfill.[135] He gave examples like the judiciary and the army. Marx and Engels observed in their class theory that these structures may be organised hierarchically, with some segments deriving more benefit than others due to their command of materials and power.[136] These relationships can be expressed thermodynamically as well: Through this artificial reorganisation of natural systems and structures, some groups of humans fulfill the function of energy sources and structures themselves rather than being full members of the fully serviced human group.[137] For instance, slaves, serfs, and, often, women, do more work and get less return out of complex societies than the owners of property and the means of production. Such disenfranchised humans are almost unable to accumulate power because slaves cannot own land and serfs cannot increase their share of the land or its product and women can often not directly inherit or manage land. Their role can be similar to the unfortunate one of beasts of burden and machines, except that the humans have similar social potential to their masters.[138]

In a thermodynamic model, the function of a slave is to produce on a very thin wedge of energy that keeps them between life and death. Sociologically and politically, the lot of serfs is to be squeezed in hard times to buffer their masters' situations.

The life expectancy of slaves and serfs is reduced compared to that of free productive land-owners. Slaves derive no benefit from their children, but the number of children the serf classes have is often greater since the only productive attribute of their situation which they may have some control over is the ability to increase the number of bodies available to work their land. We will be exploring how this situation arose in early medieval England in Book 2 of this series on the origins of the capitalist land-use system, where the British serf classes became the working classes. Note that in France, where serfs were often not allowed to marry outside their manor, fertility opportunities remained low. (See Book Three of *Demography, Territory and Law.*)

Child Labour Laws as a variable in fertility rates

Here are some explanations for changes in human fertility since the beginning of agriculture.

The demographic transition ideology is often bolstered by the assertion that people have many children where there is no social security net to look after them in their old age.

The closest I have found to theory behind this claim is that of Doepke.[139] In *Growth and Fertility in the Long Run* and in "Accounting for Fertility decline during the Transition to Growth," Doepke hypothesises that fertility falls where policies, such as education subsidies and restrictions on child labour, affect the opportunity cost of education. He compared South Korea and Brazil, the populations of which had begun to grow rapidly around the same time. He found that they differed in that South Korea had an effective public education system, and strongly enforced child-labour restrictions, whereas Brazil had a weak public education system and poorly enforced anti-child-labour laws. Doepke showed that fertility declined in association with industrialisation in Korea more than in Brazil.

To summarise, in countries where effective labour laws prohibit the employment of children, those children become costly rather than income-beneficial. In those countries where working for wages is the main option for survival for many but where child labour is prohibited, then people who rely mainly or uniquely on wages will have fewer children.

Although my own research into population growth in Britain (see Book 2 of this series) confirms Doepke's theory, there is another difference between Korea and Brazil that would have bearing on population growth. This is their respective rates of incest avoidance. Brazil has a history of a high rate of consanguinous marriages,[140] whereas Korea has an exceptionally low rate, enforced by very old traditions and modern laws.[141]

A theory of why consanguineous marriage may be important in increasing rates of marriage and therefore rates of population growth is a subject of the chapters in this book on incest avoidance and the Westermarck Effect.

Educating women

Still on the theme of education, a woman who has education will be more valuable as an income-earner than as a child-producing wife in a society that prohibits child-labour. Where women earn less than men for doing the same job, in a society which needs skilled workers and prohibits child labour, then this will be a disincentive for taking such women out of the workforce to have children. It will also be an obstacle to marriage because men's capacity to find work will be undermined by the cheaper but still skilled labour of women. In societies where monogamous marriage is the model for raising children, there are implications for marriage frequency. With children a high cost, only men with high incomes will be able to afford to take a wife out of the workforce to nurture children.

Female Land Ownership and marriage rates

In countries where men can own and inherit land, but women cannot, (England from the 12th century until the 1920s) then lack of land is an incentive for women to marry for material survival. Women who can own land and derive an income or earn a salary may experience their ability to self-support as a disincentive to marriage due to the status and power of running their own lives. A disincentive also operates in countries where, in divorce, either partner may acquire rights to the assets that the other brought to the marriage. These factors could impact on fertility rates. Some countries have facilitated the ability of women to work, raise and educate children outside marriage – e.g. France. To this should be added the fact that French women also benefit from equal inheritance rights to men. Although French women only recently (in the 1970s) regained[142] the right to manage their affairs, this right, coupled with the government's duty to house, educate and assure an income to its citizens enhances women's security and independence. France also, through its inheritance system, makes French women more likely to inherit wealth than English women, who had almost no land inheritance rights until male primogeniture was revoked in 1925.[143] Unfortunately the French situation of equity will be affected by changes to the Napoleonic Code introduced by President Sarkozi in 2008. Now it becomes possible for a spouse to make a serious claim on part of a deceased's estate where that estate previously went entirely to blood relatives and, furthermore, in the absence of children, for spouses to inherit the deceased spouse's estate, which until recently would instead have returned to the deceased spouse's parents.[144] This change is illogical because spouses already inherit from their parents (who may not disinherit them) so it is a case of depriving one family line via marriage in favour of another.

The recent ability of technological societies to prove paternity is a new factor that could be exploited to access additional income for children whose mothers might otherwise be their sole providers. This could act to increase the fertility rate, but men might become more careful about impregnating women under these new circumstances.

Abernethy's Fertility Opportunity

Anthropologist, Virginia Abernethy, theorises the presence of a 'fertility opportunity.' She has argued convincingly that where signs for the future appear positive, people are inclined to have more children. Where people perceive a slowing down of gains or a net loss – negative signs for the future - they will put off having children.[145] This is actually the opposite of the so-called 'benign demographic transition theory'. Abernethy's theory has shown a predictive ability, whereas the 'benign demographic transition' explanation of fertility lacks testability due to its absence of stated hypothesis.

Cultural Fertility Levers and the Western Blind Spot

Abernethy writes in *Population Politics*:

"Perhaps because Westerners are used to distinguishing between recreational and procreational sex, and to a pattern in which almost every young person is sexually active, we can hardly envision ways to limit births that do not rely on contraception. Overreliance on modern biological methods results in overlooking cultural and social patterns that affect a threshold factor: exposure to the risk of pregnancy.

This western blind spot can have serious consequences. For example, it encourages the assumption that modernizing will help third-world countries to control their population growth. Traditional beliefs and behaviours may be attacked simply because they are not modern. The possibility that they have had a part in limiting population growth is quite overlooked." (Virginia Abernethy)[146]

Abernethy's reiteration and analysis of classic anthropological sources on sexual and gender traditions in pre-industrial societies synthesized already what Pyrie discovered all over again in 2000.

She goes on to say that intact traditional societies, without modern contraception, usually manage to stay in balance with their resources, mostly by keeping fertility rates low. Predominantly they allow patterns to evolve which limit women's exposure to pregnancy. Rules and beliefs reinforce these patterns. "A woman who is prevented from being sexually active during most, or even all, of her adult life, will not have a large family."

Examples of patterns like this are norms of premarital virginity; gender-streaming in work, schooling and living (such as with same sex schools or in societies where men and women live in separate villages, with women raising girls and men raising boys after the age of around six). Incest avoidance and caste and other restrictions on who might marry whom are especially effective the smaller the society since opportunities increase as population increases.

The variety in conditions about who may marry whom is remarkable, with brothers traditionally sharing one wife in Tibet,[147] leaving 30% of women without marriage opportunities,[148] and, in India, the forbidding of widows remarrying, or even surviving. It has been recorded that the average age of widowhood in India rose from 29 to 35 during the 1960s,[149] creating a significantly longer fertility opportunity. Breast-feeding as a contraceptive seems to work much better if the woman is not living with the husband, as in Micronesia and other places, where women have separate houses, land and sometimes villages.[150] Lack of a dowry, property, or earning capacity reduces fertility opportunities as well.[151]

Transport

Almost never explored is the multiplier effect of opportunities to meet people outside the circle proscribed by clan and tribe that different forms of transport provide as they increase human reach beyond that of a human stride. Boats, horses, trains, automobiles, planes have been accompanied by changes in settlement scale and population numbers, as isolated and often static populations were brought into contact with others. The post World War 2 "baby boom" would not have been possible without the car. (See more relating transport to population numbers in Book 2 of this series.)

Gender-specific Fertility Levers

Gender streaming occurs in other social animals, to the extent that bears and leopards, for instance, only tend to meet up for mating, keeping separate territories and routines. However they evolved, the many often arcane-seeming but usually practical gender specific practices, of other animals and of humans, may be seen as important levers to adjust fertility opportunities up or down according to external indicators.

Life-span, Life Expectancy and Continuous Improvement

Other questions that arise in relation to theories about education costs and benefits and their relationships with fertility are the obvious ones relying on the notion of continuous improvement, especially of life expectancy. As discussed earlier in this book the assertion that absolute life-expectancy for humans has increased is highly questionable. The real story seems more to have been of an abrupt *decline* since early medieval times. Life expectancy did not return to hunter-gatherer, nomad and early medieval standards of life-expectancy until the 20th century. Nonetheless, throughout almost all periods exceptionally long-lived humans occurred. The bible's 3 score years and ten indicates that 70 years was considered a reasonable quantity to hope for more than 21 centuries ago and, as we have noted, anthropological accounts of pre-colonisation horticultural societies, such as in some Pacific Islands with their longstanding steady-state settlement indicate good nutrition and few diseases or hazards to cut life shorter than its natural limits. What most probably did impact on average life expectancy statistics where they are able to be recuperated was higher than natural infant mortality which may have been a social choice.

The modern myth about longevity improvements calls into question conventions about the life-expectancy of other animals. We generally accept that turtles, parrots, Golden Roughy fish, elephants and some other animals have life-spans similar to ours – although they have not benefited from fossil-fuel society. How wrong might we be about the life-expectancies of many other species? In the United States, they are used to cows dying aged 4, from collapsed skeletons due to merciless milking. Even in other places, a cow that makes it to 8 years is considered old; a twelve year old cow, unusually old. But some cows have made it to over 30 years old and horses to more than 50 and some perhaps over 60.[152] Some possums live about 30 years, and so do flying foxes. We are living in a world where life-spans seem to have become greatly distorted according to opportunity. The world of slaves and serfs is a world of short life-spans and most other species live in marginal habitats or are kept on marginal regimes by economically stressed farmers.

Immigration and Expected Demographic Transitions

Strangely, the general belief that modern industrial societies are participants in the demographic transition and that the demographic transition means stabilisation generally fails to take into account those industrialized countries which distort the internal demographic trends by importing vast quantities of immigrants. Examples are the United States, which has a much higher birth rate than Western Europe; and Australia, which has a very rapid population growth rate, about half of which was due to immigration from the 1960s[153] and of which the immigration component has recently doubled.

Biological Ecological Population theories

Like the philosophers and economists mentioned earlier in this short review about human population dynamics, numerous 20th and 21st century biological ecologists, through studying hunter-gatherers, have also tried to explain why they have or had steady state societies, and why the populations of agricultural and industrial societies cannot be stable. In fact, they have been wrong to think that agricultural societies could not be stable. As well as Peter Piries' examples we have already read of in Pacific Islander societies, there have been examples in Europe. Although better known for his theories about exponential population growth, Malthus also collected examples of stable populations in Continental Europe at the end of the 18th century, publishing this work in 1826. [154] Among other examples, he discussed the Swiss parish of Leyzin.

In Leyzin, a pastoral society, life-expectancy was considered by Malthus to be extraordinarily high at 61 years. Average number of the births over a period of 30 years was "almost accurately equal to the number of deaths" and emigration was not a factor or a consequence. As Malthus observed, during this period, "the resources of the parish for the support of population had remained nearly stationary." He described the pastures as limited, and not easily increased either in quantity or quality. The number of cattle, which could therefore be kept upon them was also limited and, "in the same manner the number of persons required for the care of these cattle."

Malthus theorised therefore that young Swiss men were not able to leave their fathers' houses and marry until employment as herdsman, dairymen, or similar, became vacant through a death.

> "As, from the extreme healthiness of the people, this must happen very slowly, it is evident that the majority of them must wait during a great part of their youth in their bachelor state, or run the most obvious risk of starving themselves and their families."

Malthus added that:

> "The case is still stronger than in Norway, and receives a particular precision from the circumstance of the births and deaths being so nearly equal."

Perhaps those biological ecologists are also wrong to think that industrial societies cannot be stable. As I said at the beginning of this book, there seem to be two kinds of land-use planning and population systems dominating industrial society. There is in fact good reason to think that Western Continental European societies can achieve stability, but less reason to think that the Anglophone ones can. (See Book 2 of this series for a detailed study.)

Some reasons put forward for stability in hunter-gatherer societies are prolonged breast-feeding, higher age of menarche, lower body fat storage and high rates of exercise. These factors are contrasted with differences in agricultural societies, where softer food sources were available for supplementing the diets of infants so that breast-feeding might not last for so many years, where women matured sexually earlier due to higher opportunity to accumulate fat from cereals and perhaps the exposure to oestrogens in those cereals. See, for instance, explanations for the stable numbers among traditional populations of the !Kung.[155]

Infanticide, as mentioned previously, is another factor often suggested to be important for limiting births among hunter-gatherers, but rarely is it mentioned that infanticide has also been well-known in agricultural and 19th century industrial societies[156] - which had unstable population growth rates and higher populations. Dilworth[157] does argue that infanticide and abortion were often discouraged in agricultural societies because of the need for slaves, servants, laborers and warriors, however we know that it was still practised, albeit covertly where discouraged. Speaking of agricultural societies, Dilworth also says that up to one quarter of Chinese newborns were disposed of by infanticide (he does not give dates) and that there were also very high rates in 19th century India. It is however important when referring to such catastrophic rates of population growth and the explosion of measures to combat them to look at factors disorganizing the society, rather than assume that such high rates of violent birth control were the norm for centuries.[158] In the case of India we know that massive overpopulation accompanied the disorganization of stable peoples and their territories through British colonialism. The beginning of China's population explosion coincided with Portuguese colonization of Macau in 1557 and was in full swing by 1750 as the trade wars were capped by the British industrial revolution and colonial diaspora.

The professionalisation of medicine and midwifery and the registration of births in the 20th century have made methods of aborting foetuses mysterious to the general public. Infanticide, since the mandatory registration of births at the beginning of the 20th century, has become a highly detectable form of murder. Documents indicate that, when women themselves and local midwives more or less exclusively dealt with pregnancies and birth themselves, abortion and infanticide were more current, if covert.[159]

Most of the references I have used here for hunter-gatherers use the !Kung as important examples.[160] They also refer to the kinship rules of the !Kung which forbid marriage to people with particular names that signify something like the 'moiety' kinship divisions as well as to first, second - and possibly third cousins, aunts and uncles - both blood relatives and in-laws. Despite this knowledge, there is widespread failure to make the connection between reduced fertility opportunities and stable populations. References do mention and try to account for the small, well-spaced families, which would not depend on marriage opportunities, but they do not look at the fact that women often go back home to live after getting married, and that men and women spend large periods of time apart due to their different gender roles - the women gathering and the men hunting. References also mention that the !Kung lived with their huts very close together and could therefore easily prevent woman or child abuse, but there is no mention of the impact of the Westermarck Effect, which one would assume to be widespread and profound given the continuous proximity of clan members. The next chapter will discuss the Westermarck Effect in some depth.

Kaplan et al 2009 in their Evolutionary Theory of Human life span (discussed earlier) establish the validity of considering a theoretical genetic basis for any durable animal characteristic. The theory I will advance in the next chapter is not about how our genes interacted with our environments over millions of years to give us big brains or long life spans, but about how DNA may interact with social and physical environment in 'real time' to regulate sexual interest and fertility. It is a new theory of Incest avoidance and the Westermarck effect as part of a genetic algorithm[161] for population size affecting many lifeforms.

CHAPTER 3: THE URGE TO DISPERSE: WHY CHILDREN DON'T USUALLY MARRY THEIR PARENTS

This is a new theory on how incest avoidance and the Westermarck Effect (which you will learn about) exert an effect on the pattern of human settlement and the numbers of humans on earth. It is not a theory to discredit the related implications of genetic diversity, but it argues that avoidance of 'inbreeding', as well as promoting genetic diversity, also causes population spacing and limits fertility opportunities.

The *Urge to Disperse* theory[162] that follows derives from examples of how dispersal norms also affect numbers in other animal and plant populations and explains how such populations are usually able to remain stable within local environmental limits.

In what might to the non-anthropologist seem a rather startling leap, the Urge to Disperse theory identifies a way in which such innate patterns ultimately contribute to human political organization in the form of inheritance and land-use (property ownership) traditions and laws. These new-sounding ideas actually have their basis in well established anthropological norms, which are discussed below these introductory paragraphs. It was my curiosity about the political implications of the effects of these patterns in human and ecological history that inspired *Demography, Territory and Law* and which has permitted some radical new explanations for how we all got to where we are now in the other three volumes. The implications are that by preserving natural organization of populations within a landscape and preserving local economies, steady-state populations should be achievable. This contradicts the message that overpopulation is an inevitable stage that must be borne in every society.

The initial title of this chapter was, "The Urge to Disperse: Towards a new theory on Function of Incest Avoidance and the Westermarck Effect in a biological algorithm for population spacing." Since the word, 'algorithm' can cause indigestion through both fear of pretension and fear of impenetrable concepts, I changed the title to make it more immediately meaningful. Nevertheless, algorithm seems to be the best term to apply here, so what is a biological 'algorithm'?

A biological algorithm is an inherited response tailored to an internal or external situation in an animal or plant. The term has been used well by E.O. Wilson,[163] who wrote,

"Instinctive habitat selection is universal, and its analysis has become an important industry within the growing discipline of behavioral ecology. Even tiny species of insects, rotifers, and other invertebrates, whose brains are invisible to the naked eye, follow impressive algorithms of orientation to reach the habitats they need to survive."

He also used the term, 'epigenetic rule', "defined as an inherited regularity of development." And he gave incest avoidance and the Westermarck effects as examples of epigenetic rules.

"Westermarck effect is an example of an epigenetic rule, defined as an inherited regularity of development."

Incest avoidance and the Westermarck effect as indicators of biological algorithms for population dispersal, spacing and size, in many species including humans

The next two chapters look at the role of incest avoidance and the Westermarck effect as evidence of biological algorithms for population spacing and dispersal patterns which, if undisturbed, would keep populations naturally within carrying capacity. It also explains why populations may overshoot or fail to establish. The implications are, broadly, for maintenance of steady state populations – human and others - within a rich and diverse ecology. The theory uses both social and biological science.

Review of the Literature (Social and Zoological)

Social Theory:

All societies have laws relating to incest, overwhelmingly banning it as 'taboo', with some exceptions, usually in ruling castes, famously with the Egyptian pharaohs, where it was actually prescribed.

Levi-Strauss: Incest Prohibition as the Socio-sexual Organiser

The theory of incest avoidance as a major ordering force of population spacing is a well-known anthropological one which has been written about by, among others, Levi-Strauss. The theory is that the sexual and the social intersect dynamically. Strauss identified the prohibition of incest as the fundamental socio-sexual 'organiser'. Prohibition of incest – that is to say, of sexual relations with blood-relatives or with the same 'class' (e.g. relatives by marriage) – acts universally as the point of encounter where the sexual meets the social. In *Elementary structures of families*, (1949) (*Les Structures élémentaires de la parenté*, 1949), Claude Levi-Strauss interpreted this as the decisive moment of the passing of nature into culture, and as the pivot of a movement that required the exchange of women. From this point, he argued, stems the organised system of marriage based alliances and inheritance – that is, the social field.) A British ethnologist, Robin Fox, treated the problem from an evolutionary perspective: Groups with very strong drives would have established moderating mechanisms, such as the prohibition of incest, which would have led to their 'adaptation'. (*Anthropologie de la parenté*, 1967). [164]

The explanation above assumes that incest avoidance was the result of prohibitions, but, as I will elaborate below, zoological observation and testing of hypotheses indicates that incest avoidance is present in other species, where conscious prohibition is not used as an explanation. Tests reveal suppression of ovulation and testosterone production in the presence of close relatives in non-human species.[165] This is a reason why I prefer the term, "incest avoidance" over the term "incest prohibition".

Biological theory:

What is the biological function of the almost universal taboo, also known as 'incest avoidance'?

Most people think that they know, and that statement includes psychiatrists, zoologists and sociologists. The usual explanation is that the incest taboo is a social response to the costs of incest. It is inferred by sociologists that societies noticed that incestuous relations gave rise to unhealthy children.[166] Although it is true that close inbreeding increases the risk of inherited diseases, fault may be found with this explanation in that many other creatures practice incest avoidance as assiduously as humans. Examples include tiny and presumably unscientific organisms, such as cockroaches, which we would not expect to make conscious decisions on the grounds of observation. But, perhaps we underestimate cockroaches.

Also against the culturally mediated genetic diversity impulse explanation is the application of incest taboos to 'in-laws', in a more or less symmetrical fashion to true incest avoidance. By breeding with in-laws (who are not blood related) the gene-pool would increase, but this opportunity is avoided and further dispersal remains necessary to find mates. In fact in-laws are in a sense indirectly blood-related by marriage, and potential producers of directly blood-related offspring.

Working within the progress ideology, Anthropologist Levi-Strauss (1908-2009), rejected biological explanations of human social behaviour. Unaware of the occurrence of incest avoidance in so-called primitive societies and in other species, he believed that incest avoidance was a cultural evolution from savagery to civilized society.[167] Marx also theorized this. Freud believed that culture had invented the incest taboo in order to mitigate an extremely strong drive to incest in humans. Levi-Strauss believed that humans attenuated hostility between groups and encouraged alliances by having their children marry outside their own group. He also thought that it would become too complex and difficult to identify the status of a group's members if they were all very closely interrelated. He thought the outcome might be for the family unit to break down under the stress of sexual and territorial conflicts. Those explanations are still reasonably useful to explain an outcome of incest avoidance, which is social organization.[168]

The seeking of mates within one's own group is called 'endogamy'. Dispersal to find mates outside one's own group is called 'exogamy'. Traditionally it has worked within a restricted continuum, between clans and tribes, where clan members found marital partners among more distantly related clanspeople, but still part of the same tribe.[169]

Interestingly, Freud was inspired to his theory of incestuous impulse (*oedipal conflict*) by his interpretation of the story of *Oedipus Rex* in Greek mythology. Ironically, the Greeks actually got the mechanism right, for Oedipus was only able to be attracted to his mother because he was not raised by her. (See further on for the Westermarck effect. Freud's theory had only retained the fact of an incestuous attraction. Freud thence theorized the presence of an 'oedipal complex' with its feminine the 'Electra complex' as a part of human development in order to explain the frequency of reports of incest from his patients, which he decided were most probably fantasies.

Freud was unaware of the Westermarck effect, (see below) however this effect might have provided him with another explanation for the rate of incest reports he experienced in late 19th century Vienna. In an era and social strata, in households where nurses took over the mother's role and fathers often spent much of their time away from home at work, or otherwise socially distanced from their children, the Westermarck effect may not have been effective. If that was so, then the barriers to incest between father and daughter, mother and son, might have been reduced.

In modern industrial society, the pattern where both parents work long hours away from home and children, may carry a similar danger.

The Westermarck effect

Towards the end of the 19th century, Finnish sociologist, Edvard Westermarck, whilst conducting research into endogamy and exogamy, discovered a phenomenon which came to be called the *Westermarck effect*. He was able to show that incest avoidance applied to people raised together, whether or not they were genetically related. The effect has since been observed many times over, notably in a longitudinal study of kibbutzes by Shepher in 1983.[170] In this study of children raised communally in peer groups, of the 3,000 marriages of those children in later life, only 14 were to members of the person's childhood kibbutz peer-group, and none of those 14 had been reared together during the first six years of life.

The absence of the Westermarck effect coincides with the absence of incest avoidance responses where close blood relatives are raised apart from each other.

What does this mean? Does it mean that nature is easily fooled about blood relationships and causes meaningless oppositions but allows dangerous attractions?

If we look at incest avoidance and the Westermarck effect more broadly - as impulses for population dispersal rather than moral or social arbiters - then the mirror phenomenon of blood-relation avoidance, the adoptive-relation avoidance, and the indirect blood-relation avoidance (in-law avoidance) seem to more consistent and coherent. I don't think that we can, however, diminish the structural importance of incest-avoidance, since this seems clearly to set the initial pattern of population dispersal in time and space.

Although mainstream sociology has never really embraced the theory of a biological basis to incest avoidance, research has progressed in zoology, leaving Levi-Strauss, Freud and Marx behind, but not Westermarck. That is to say that the Westermarck Effect answers a lot of questions, however the phenomenon does not appear to be widely known among zoologists.

Incest avoidance and the Westermarck effect as a population dispersal mechanism affecting population size and political organization through inheritance of territorial rights

In the preceding chapter I reviewed a number of theories about human population growth. This chapter looks at research and theory into the incidence and impact on social structure of incest avoidance, the Westermarck effect, and inheritance patterns. It advances a new theory of incest avoidance and the Westermarck effect in population dispersal as an independent variable of land-tenure, pattern of land-settlement, population size and density. It hypothesises that the way human societies are organised depends greatly on how they link land-use to inheritance, which in turn links the land to the social structure of the family, clan, tribe and village. Also in this chapter are new suggestions about the effect that disruption of incest avoidance/Westermarck effect and their relation to place could be having on modern populations in aggregate and on individuals. These chapters do not focus on incest itself and especially not in the social sense of a criminal or depraved act. This is not a psychiatric or a criminological investigation. It is also not a discussion about the risks of inherited disease or deformity due to inbreeding. Essentially it is about how the need to go away from the primary family unit to find a sexual partner is both a limiter of reproductive opportunity and a primary impulse for population dispersal.

After these chapters the theory will be used to reinterpret early, medieval and modern history with particular reference to theories of societal collapse and the development of expansive and conservative economies and populations.

Identifying a New Function of Incest-Avoidance and the Westermarck effect: Population Dispersal

To state what may seem obvious and trivial: incest avoidance would be almost automatically guaranteed in large populations due to the laws of probability.

Conversely, in small, isolated populations incest is very likely without incest avoidance strategies and unavoidable under normal conditions of fertility for communities under a certain size to avoid extinction. In such cases exceptions to the incest avoidance rule may or may not occur. Where these exceptions occur they cause caste systems in humans, sub-species in other animals, initial territorial aggregation and finally overshoot. (See further on for examples under the heading, "Isolated Populations.")

In the preceding chapter I showed that incest avoidance laws are very important components of the Pacific Islander land-tenure system, which is a system that occurs in relatively isolated, small populations, but equivalent systems almost certainly underpin every other society on this planet.

In this chapter I want to offer material to support the idea that incest avoidance and the Westermarck effect may be part of a response to genetic algorithms which underpin the organisation of human settlement; i.e. that incest avoidance is an instinctive norm in humans.

Incest avoidance and the Westermarck effect and the related population dispersal seems to be a dynamic deeply embedded in most or all sexually reproducing species, even including the reluctance of hermaphrodite plants to 'self' themselves.

The impulsion for population dispersal saves species from isolation and consequent annihilation through local accidents. It also fits the 'selfish gene' theory of Dawkins,[171] by promoting maximum survival opportunity for genetic material in a multitude of life-forms in many different places. It is also logical and practical, or coral chains and human settlements would (oxygen, food, and musculoskeletal structures permitting) reach up vertically to the moon and beyond.

Numerous studies, some of which are discussed in this chapter, demonstrate that incest avoidance occurs in many species. Examples discussed here include the acorn woodpecker, the superb fairy wren, voles, mice, butterflies, marmosets and primates and could have included toads, cockroaches and more.[172] For instance, in hermaphrodite species, including plants:

> "Another type of inbreeding avoidance is avoidance of selfing in hermaphrodite species. This is of great interest in particular because low selfing rates or even complete selfing avoidance are very commonly found in plants, and also in most hermaphrodite animal taxa.[173] Avoidance of selfing can be expected to have the same effect in our model as sibmating avoidance in dioecious species. This is because from the cytoplasm's viewpoint, a hermaphrodite is essentially the same as a pair of siblings in the context of reproduction. Moreover, at least one of the mechanisms of selfing avoidance—self-incompatibility—can be expected to also lead to incompatibility between closely related individuals, thus resulting in even higher outbreeding rates than with mere selfing avoidance."[174]

Human land-tenure systems as outgrowths of biological algorithms

Some of the research published shows clearly that incest avoidance impacts on population spacing (i.e. density and fertility), even though the research has been done for different reasons – often to elucidate scientific or political arguments about whether incest avoidance is genetically or culturally based.

Although my interest is not in the political polemics of nature vs nurture, the politics of the nature versus nurture argument regarding the origins of incest avoidance invite consideration of the role of culture in land-tenure patterns and of biology in culture to me. How densely we occupy land is a function of population spacing. The pattern in which populations disperse and organize in space is initiated by incest avoidance and the Westermarck effect. These observations will later underpin a theory to explore and explain questions about human population overshoot and different political systems and ways of distributing energy.

Thus our human land-tenure systems seem to be outgrowths of biologically based algorithms. Work done on the socio-sexual organization of birds, monkeys, apes and other animals, gives reason to suppose that a system of hormonal feedback from the environment probably informs the application of genetic algorithms to local conditions.

Population Numbers and Regulation

Some of this work has been done to evolve theories as to why some populations overshoot their resource base and others do not.

One such theory is the 'mechanistic' predator-prey theory, which hypothesises that some or many populations may be kept in check simply by being slain and eaten by predators. Dependence of the predators on the predated population in turn keeps the predators in check. In its crudest expression this theory ignores in-migration, out-migration, and fails to establish whether the initial population arose naturally or chaotically, which impacts on rate of incest avoidance and the Westermarck effect, often supposing that all populations simply breed as much as they can, if they do not benefit from modern contraception, albeit with seasonal variations.[175]

A more sophisticated appraisal of what may be involved in population regulation is to be found in Pimentel, "Population Regulation and Genetic Feedback".[176] The author identifies a number of rules. One is that most species are quite rare, relatively or 'by whatever criterion they are judged'.[177] This rule helps to construct the idea that huge numbers involved in overshoot by a species are probably rare and do not last for long. Another is that nearly all animals feed off live material. This observation is important because dead material cannot evolve genetically in response to predation. Pimentel describes field observations and laboratory tests which show that predated populations evolve in response to a particular predator "only if the numbers of the animal are sufficient to exert some selective pressure on the host." Using a variety of examples, he observes that the dominant control mechanism operating initially is "competition" (meaning selection), "but genetic feedback became dominant with time and through evolution." He observes that "subtle genetic changes" affect the predator, and gives this example:

> "For instance, when young pea aphids (*Acyrthosiphum pisum*) were placed on a common crop variety of alfalfa (*Medicago sativa*), they produced a mean of 290 offspring in 10 days, whereas the same number of aphids for a similar period on a resistant alfalfa variety produced a mean of only two offspring. In another example, the mean rate of oviposition (eggs per generation) of the chinch bug (*Blissus leucopterus*) on a susceptible strain of sorghum (*Sorghum vulgare*) was about 100, whereas on a resistant strain the mean oviposition was less than one. In both, reproduction in the animals feeding on the resistant plant hosts decreased more than 99 per cent. This reduced reproduction obviously would have dramatic effects on the population dynamics of the feeding animal populations."[178]

In Hopfenberg and Pimentel, "Human Population Numbers as a Function of Food Supply,"[179] the authors cite evidence of other species decreasing their fertility by a variety of means which would suggest a hormonal feedback mechanism from food availability in the environment.

> "Some species self-regulate their number to their food resources by maintaining home ranges. Chitty (1995) reported that excess young voles, for example, are forced to leave the home range of their parents... . Possibly more germane is the evidence that a sudden improvement of diet in sheep causes an increased ovulation rate (Schinkel, 1963) and that fasting in mice for relatively short periods of time

prior to mating resulted in depression of male libido and reduced conception in females (Christian et al., 1965)."[180]

They attest that this simple relationship between food supply and fertility is also seen in hunter-gatherers who do not have massive infrastructure to complicate their interactions with the environment.

"The populations of human cultures described as hunter-gatherers were limited to the food resources available (Lee, 1969; Lee and DeVore, 1976; Pimentel and Pimentel, 1996). Where these cultures still exist untouched, this continues to hold true."[181]

Biologists know that population density itself is a major indication of soil and climate fertility.[182]

Need to consider a different kind of mechanism

These studies and observations, however, do not identify a mechanism which would limit fertility in order to avoid indefinitely the experience of restricted calories. If food shortage were required before fertility dropped then clans and herds would be permanently on the verge of starvation. That kind of stress would not be conducive to genetic survival.

"... Iwamoto (1978) has shown that monkey troop size increases rapidly after artificial provisioning, but the level of consumption efficiency of the troop is always maintained lower than the critical point in both the artificial and natural habitat. Starvation within the troop simply does not occur if the rate of food availability is held relatively constant." [183]

There has to be another mechanism whereby species can adjust collectively to local environmental constraints of their ranges without constantly risking starvation.[184]

As early as 1940, in "Australia: Ecology, spacing mechanisms and adaptive behaviour in aboriginal land tenure", Joseph B. Birdsell's meticulous study of marriage laws and patterns in Australian aborigines over a wide range of climates showed that strictness of incest avoidance strongly corresponded with environmental fertility as measured by rainfall.[185] Naturally soil quality and quantity and temperature were other quotients but rainfall was the indicator that Birdsell measured. We can see that rainfall would be simpler to quantify for such research and possibly a more reliably distributed indicator than soil or temperature range. In these regions it is actually the decisive factor in environmental fertility.

Which are likely indicators of environmental fertility that may mediate fertility and space in humans and other species?

From Birdsell's observations above, it seems that rainfall [plus soil] is probably an indicator of environmental fertility which could be objectively measured in food availability and that potential, average and seasonal food availability restricts animal fertility in some way well before the point of food shortage. Birdsell also identifies (in humans) the fact that the social mechanism (also present in other animals) of incest avoidance limits fertility opportunities and can be varied in strictness in response to conditions.

What might be a mechanism whereby incest avoidance adjusts to environmental variations over regions?

It does not seem likely that the many species of animals which practice incest avoidance make complex conscious calculations about carrying capacity and then respond by devising such codes of conduct. More probable is that hormones are a mediating factor between reliability of food availability and fertility. They might be thought of as a chemical messenger.

Living creatures are biochemical systems which are regulated by hormones. Not all hormones are concerned with sexual reproduction, of course. Hormones regulate absolutely all our functions, from appetite, sugar uptake and digestion, to sweating, tissue growth, body shape, muscle quality, and sexual maturity and sexual behaviour. Hormones are responsive to external and internal environmental conditions. Some hormonal responses are obvious and well known, such as the impact that Spring has on reproduction in most living things – plant and animal – and the response of sexual attraction that occurs on encountering a prospective mate.

Studies in animals and in people show that hormones are also affected by the presence of close family. One of these effects seems to be the suppression of oestrus in incest avoidance. Relevant studies on acorn woodpeckers, marmosets and apes are discussed in this context further on.

Based on the role of hormones described above and the objective data from the animal studies below, my hypothesis or answer to the questions I have framed above is that hormones will deliver more or less fertility according to availability of living space. Space (territory) required per individual will be affected by density and reliability of food distribution, and all of this will be mediated by some degree of incest avoidance/Westermarck effect, which is also related to social dominance. (Other forms of dominance such as caste and gender may also operate. Opportunities for meeting may also be limited by gender specific circuits and territories, which can themselves be seasonal.)

A minimum physical distance from blood relatives or 'in-laws' would be required before sexual activity becomes likely. If a subject who is of an age to reproduce is dominated (i.e. within the physical sway of a parent) ovulation and sexual maturity may actually be measurably chemically inhibited.[186]

In some species sexual maturity or ovulation is delayed or permanently postponed where not enough territory is available. Where ovulation does not occur in females, even though males may have viable testosterone levels, their sexual activity will be decreased or absent. This is presumably related to opportunity and opportunity is in part defined by the presence of oestrus (being in season) and oestrus is limited by the close presence of relatives.[187]

Studies of Incest avoidance in other species

Biologist and ecologist, E.O. Wilson, reports in "Nature Matters" on the high incidence of incest avoidance algorithms in many species which have been detected in biological anthropology, sociobiology, cognitive psychology and neuroscience studies:

> "These algorithms can be blocked or reversed only at the peril of mental health. An example is the negative imprinting that forms the basis of incest avoidance, as follows: When either of two persons lives in close domestic proximity during the first 30 months' life of either one, both are unable to form close sexual bonding later in their lives. The phenomenon, known as the Westermarck effect in honour of the Finnish anthropologist who discovered it a

century ago, is evidently widespread if not universal in human beings. Equally impressive, it is shared by all other primate species whose sexual behavioural development has been closely studied.

The nature of the automatic incest-avoidance process, as well as its evolutionary origins, now seems well established by solid research."[188]

He concludes that there is a clear adaptive advantage "to those who react to it correctly". He believes this is the avoidance of "inbreeding, fewer homozygous defective genes, and more healthy children". He writes that the "*Westermarck effect is an example of an epigenetic rule, defined as an inherited regularity of development.*"

Whilst I agree that incest avoidance promotes genetic diversity and thereby reduces the disease risks of inbreeding, as I have already mentioned, my own conclusion is that population spacing is the most important effect. Inbreeding does carry risks but population spacing is absolutely vital to social organisation.

Acorn Woodpeckers – incest avoidance in a complex social order

Walt Koenig and Joey Haydock, in the "Social Behavior of the Cooperatively Breeding Acorn Woodpecker,"[189] give a riveting report on a study which has been going on since 1971. Their observations show that 'mere' birds maintain symmetrical incest avoidance in the most complex of social orders. The reader may draw the conclusion as I do that incest avoidance is the basis upon which this social order is built. The study shows that sexual maturity does not occur without the presence of suitors who are genetically sufficiently distant. This implies complex hormonal suppression. Specific hormone suppression is actually tested and measured in a study following this one.

"Acorn Woodpeckers are common residents of oak woodlands in western North America.

Groups engage in a many communal activities, including territorial defence, feeding of young at the communal nest and acorn storage in special trees known as granaries. Stored acorns are an important food resource, both during the winter and for successful reproduction the following spring. Groups can even breed in the fall when the acorn crop is particularly good.

They have one of the most bizarre mating systems of any bird in the world. They are co-operative breeders and live in groups composed of up to six co-operative breeder males, three joint-nesting females, and non-breeding helpers of both sexes.

Co-breeder males are brothers and/or fathers and their sons competing for matings with the joint-nesting females, who are sisters or a mother and her daughter who lay their eggs in the same nest cavity. Offspring produced from this communal nest may remain in their natal group for several years as non-breeding helpers, during which time they help feed younger siblings at subsequent nests.

This kind of mating system is known as polygyandry. All individuals within the group are close relatives except that co-breeder males are not related to joint-nesting females. Incest avoidance is maintained because helpers only inherit and become co-breeders following reproductive vacancies when the breeders of the opposite sex die and are replaced by unrelated birds from elsewhere. Reproductive vacancies are often filled by a unisexual set of siblings who compete against other sibling groups in spectacular events called power struggles. Winners of power struggles become co-breeders in the new group; losers return home and resume non-breeding helper status. (...)"

The mathematical and spatial consequences of these practices would be a reduction in fertility opportunities which would otherwise be available if any male could mate with any female. Within the prevailing arrangements more territory is available per breeding group than would be available if most birds bred.

As odd as the Acorn Woodpecker's households may seem to us, they may not be all that unusual among birds. The Superb Fairy Wren was recently the subject of a study which demonstrated very similar arrangements in Cockburn, Osmond, Mulder, Green and Double, "Divorce, dispersal and incest avoidance in the cooperatively breeding superb fairy-wren, *Malurus cyaneus*".[190]

I have quoted their useful numbered summary below.

"1. Between 1988 and 2001, we studied social relationships in the superb fairy-wren Malurus cyaneus (Latham), a cooperative breeder with male helpers in which extra-group fertilizations are more common than within-pair fertilizations.

2. Unlike other fairy-wren species, females never bred on their natal territory. First-year females dispersed either directly from their natal territory to a breeding vacancy or to a foreign 'staging-post' territory where they spent their first winter as a subordinate. Females dispersing to a foreign territory settled in larger groups. Females on foreign territories inherited the territory if the dominant female died, and were sometimes able to split the territory into two by pairing with a helper male. However, most dispersed again to obtain a vacancy.

3. Females dispersing from a staging post usually gained a neighbouring vacancy, but females gaining a vacancy directly from their natal territory travelled further, perhaps to avoid pairing or mating with related males.

4. Females frequently divorced their partner, although the majority of relationships were terminated by the death of one of the pair. If death did not intervene, one-third of pairings were terminated by female-initiated divorce within 1000 days.

5. Three divorce syndromes were recognized. First, females that failed to obtain a preferred territory moved to territories with more helpers. Secondly, females that became paired to their sons when their partner died usually divorced away from them. Thirdly, females that have been in a long relationship divorce once a son has gained the senior helper position.

6. Dispersal to avoid pairing with sons is consistent with incest avoidance. However, there may be two additional benefits. Mothers do not mate with their sons, so dispersal by the mother liberates her sons to compete for within-group matings. Further, divorcing once their son has become a breeder or a senior helper allows the female to start sons in a queue for dominance on another territory. Females that do not take this option face constraints on their ability to recruit more sons into the local neighbourhood."

This above should make it clear that the practice of females travelling to avoid incest would profoundly affect the spatial patterns of settlement and the rate of population growth.

The cooperative breeding might also improve the chances of fledglings reaching adulthood – and thence raise the population growth rate - although this is a moot point since, at least in the case of the Acorn Woodpeckers, nesting female birds frequently got rid of their sisters' eggs from the nest."

Incest Avoidance in Non-human Primates

In "Constraints on control: factors influencing reproductive success in male mandrills (Mandrillus sphinx)",[191] Marie Charpentier, Patricia Peignot, Martine Hossaert-McKey, Olivier Gimenez, Joanna M. Setchell, and E. Jean Wickings, write that the *"Mechanisms of inbreeding avoidance are well documented in vertebrate societies"*, and cite a number of well established references.[192]

They inform us that the social structure of most primates, including macaques and the mandrills in this study, is such that the males tend to disperse after adolescence to distant groups, whereas the females remain mainly with the home group.[193] (The opposite occurs with chimpanzees and is more common in human dispersal.)[194] They confirm that in the mandrills they studied, social rank is inherited by females via a female line, but males generally leave the natal group to seek positions in other groups.

Their study showed that incest avoidance was marked where there was 50% relatedness (as in siblings or parents to children), and that incest avoidance decreased as the blood relatedness decreased. The subjects were observed in a colony that was artificially isolated, so that the male tradition of joining distant colonies could not be carried out. Tests showed that males who were obliged to remain with the colony also observed incest avoidance, which demonstrated a knowledge of kinship, according to the authors.

"Several studies have shown that maternal relatives avoid mating with one another (rhesus macaques: Smith, 1995; red colobus, Procolobus badius temminckii: Starin, 2001; Japanese macaques: Takahata et al., 2002; and see for review: Moore, 1993; van Noordwijk and van Schaik, 2004), but less is known concerning patterns of inbreeding avoidance between paternal relatives (but see Alberts, 1999). In this study, we showed that the probability of paternity by a dominant male decreased when he was related to the dam at R = .5 (the highest possible relatedness coefficient in our study). Smith (1995) showed in rhesus macaques that the intensity of inbreeding avoidance was directly correlated with the closeness of kinship, as in the mandrills studied here."

The authors also write:

"The incest avoidance shown here demonstrates that female mandrills may exercise an active choice of partner, avoiding mating with close relatives, as raising an inbred offspring may be costly. A detailed study of female mate choice is indicated to determine the effects of relatedness."

(Obviously I don't agree with the inference that females avoid incest in order to avoid genetic disease problems).

"Surprisingly, in this study, we also showed that the more closely the dominant male was related to adolescent males in the group, the higher was his probability of paternity. This is the opposite of the finding concerning adult males. In the mandrill colony, different mating strategies appear to be employed at different times during a subordinate male's lifespan. A hypothetical pattern could be proposed. During adolescence, males appear to support a closely related dominant male and hence may avoid aggressive interactions that they cannot win, but on reaching adulthood, the situation is reversed and subordinate males may compete successfully for females, dominant males being less vigilant with closely related subordinates."

My comment is that this also shows kinship and territory relationships. It also resembles the strategy adopted by young human males towards their fathers and could provide clues to an evolutionary political explanation for the inheritance tradition of primogeniture, which will be discussed in the next chapter. The statement that "the more closely the dominant male was related to adolescent males in the group, the higher was [the dominant male's] probability of paternity" is an indication to me of lower sexual activity in male offspring on the parental home range, except of course in the father or other dominant relative.

Marmosets and Hormonal Suppression in the presence of dominant close family

"Common marmosets live in social groups in which all group members help to raise the babies of a single dominant female. In marmoset groups only the dominant male and female breed whereas lower ranking or subordinate marmosets do not because their reproduction is repressed."[195]

In this report entitled "The Common Marmoset, Callithrix jacchus, Current Research",[196] sexual immaturity or infertility in the presence of dominant close family is actually linked to suppression of oestrus in females, with measurably smaller ovaries. The infertility in female marmosets is similarly affected by the presence of non-related dominant females. (Note that, in humans, girls raised in the presence of their blood related fathers reach menarche (i.e. menstruate for the first time) later than girls raised in the presence of a non-blood related father.)[197]

Male marmosets have reduced rates of sexual activity in the presence of dominant close family where the chances of incest are high. Their testosterone levels, however, are not affected. This may indicate a behavioural avoidance of incest, but it could be due to the lack of available females in season, which seems itself to be a response to being unable to find enough territory to escape parental presence or the more dominant unrelated females. This behavioural choice supports the Marmoset family structure where only one couple breeds and the other relatives assist in the raising of that couple's babies.

Here is an extract from this interesting report.

"Reproductive Constraints:

Subordinate or low ranking male and female marmosets do not reproduce, however researchers believe it is because of several reasons. Subordinate males and females appear not to breed because of the risk of incest and behavioural limitations whereas subordinate females also have hormonal constraints. Common marmosets live in social groups in which all group members help to raise the babies of a single dominant female. In marmoset groups only the dominant male and female breed whereas lower ranking or subordinate marmosets do not because their reproduction is repressed.

This suppression of reproduction is the case in many primate and non-primate mammals including yellow baboons. Dominant marmosets, unlike dominants in other species, however, do not appear to use harassment as a means to keep the subordinates from breeding.

Researchers have shown that the ovaries of the subordinate female marmosets are about half the size of the dominant females. Blood samples taken from subordinates to detail hormone levels also revealed that subordinates do not ovulate. When the subordinate females are taken out of the presence of a dominant female and placed on their own they will ovulate. Researchers, therefore, believe that reproductive suppression in subordinate females is due to the hormones of the subordinate female. One hormone believed to be involved is released from the brain and stops the release of hormones stored in reproductive organs. When researchers gave large doses of this hormone to subordinate female marmosets they began to ovulate and when the levels of the hormone were reduced the female stopped ovulating.

Researchers believe the lack of this particular hormone has something to do with the reproductive suppression of subordinate female marmosets.

Researchers have not pinpointed which cues, visual, behavioral, or olfactory (smell), produced by dominant female marmosets cause subordinates to stop ovulating and, in turn, stop reproducing.

Like the subordinate female common marmosets, subordinate males also do not reproduce.

Researchers, therefore, wanted to know the causes of male reproductive suppression. Researchers thought that reproductive suppression of subordinate males was due to cues given by dominant males or as an incest avoidance behaviour to keep family members from mating. For the study subordinate males and their fathers were tested either alone or together in a cage and were joined by another female familiar to them (like their mother or mate) or an unrelated and non-familiar female. The number of sexual behaviours the males engaged in was recorded and blood samples were taken to measure hormone levels. Researchers found that the sons engaged in very low rates of sexual behaviour with familiar females whereas their fathers (the dominant male) engaged in higher levels. This might imply that the subordinate male does not mate with familiar females because of the chance of being related. When fathers and sons, however, were tested individually with unrelated females, both engaged in approximately the same number of sexual behaviours.

This would suggest that subordinate male marmosets have no problem mating if it is with a non-familiar female. Finally, the hormone levels were examined and the researchers found no difference between the levels of fathers and sons. Researchers think that because subordinate male marmosets do not engage in much sexual behaviour with familiar females and because hormone levels are similar between dominant and subordinate males that subordinate males are reproductively suppressed to avoid incest.

Researchers now have evidence to believe that subordinate male common marmosets do not reproduce so as to avoid incest. Subordinate females, however, are thought to be reproductively suppressed because of hormones. Presently, researchers are still studying the cause of reproductive suppression in male and female common marmosets."

These zoological reports demonstrate the presence of incest avoidance in several social species. The practice of incest avoidance reduces potential population size and density, providing more territory per animal than would otherwise be available. Without incest avoidance, populations would reach much higher density, reducing average territory per capita. Weight, size, quality of life and longevity would suffer and one would expect increased aggression.

We can also see political behaviours accompanying incest avoidance, such as submission or harassment (in baboons, for instance). In marmosets and cooperatively breeding birds, there is a politico-social adaptation. These organisational patterns seem to be algorithms with incest avoidance and the Westermarck effect an obvious denominator.

CHAPTER 4: TOWARDS A NEW SOCIAL THEORY ON POPULATION DENSITY AND GEOMETRIC PATTERNING

How densely we occupy land is a function of population spacing. The pattern in which populations disperse and organize in space is a function of incest avoidance. For this reason, some basic and fractal connection between human land-tenure systems and political economies and instinctive population dispersal may be inferred. These observations will underpin a theory to explore and explain questions about human population overshoot and different political systems and ways of distributing energy.

For the traditional sociologist who is trained not to compare humans to other species, this line of thinking may seem hopelessly radical. Remember, however, that it has long been a sociological tradition to jealously claim incest avoidance as something which distinguishes humans from 'lower' creatures. Do we now reject discussion of this phenomenon because it binds us so irrevocably to the rest of the animal kingdom? Or do we accept that incest avoidance as a dynamic in population spacing implies that there is potential for cooperation rather than for conflict?

Let us now proceed as if incest avoidance – both blood and 'in-law' - or population dispersal along principles evoked in the Westermarck effect - is the norm.

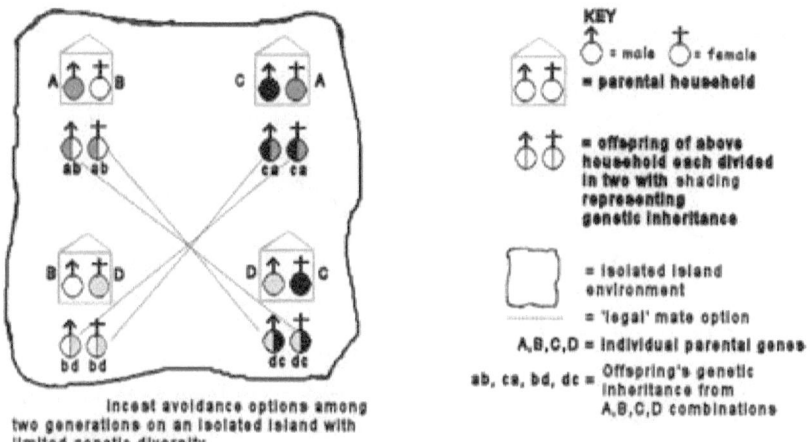

Incest avoidance options among two generations on an isolated island with limited genetic diversity.

Figure 6. Incest avoidance options on isolated island

If we think about the patterns that result across a landscape and in time from incest avoidance the results would be a binary system that grows in symmetrical geometric progression, according to landscape fertility and time. It would be organized fractally. This binary system defines the limits of fertility opportunity according to the norm of incest avoidance. In such a system, brother may not marry his sister, daughter or mother. Usually brother may also not marry grandmother, aunt or cousin or niece. Of course this also means that sister may not marry brother, son, father, grandfather, uncle, cousin or nephew. Siblings of the same two parents are related to each other at the rate of 50% each. Cousins, grandparents, nieces and nephews, aunts and uncles, are related by 25%.

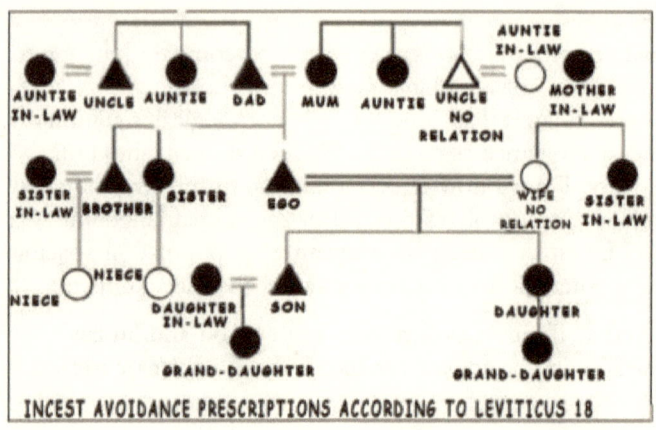

Figure 7. Incest avoidance prescriptions – Leviticus 18[198]

Figure 7, "Incest avoidance prescriptions according to *Leviticus 18,*" is a geometric illustration of incest avoidance prescriptions according to *Leviticus 18* in the bible. 'Ego' is the central reference person. Patrilineally-related men are shown as triangles; excluded marriage partners are shown as dark circles; partners not explicitly excluded are shown as lighter circles.

The illustration above gives a good idea of the reduced range of opportunities to sire legal heirs in a typical, well-known and well-documented incest prohibition system, usefully documented in Leviticus 18 in the Bible.[199]   Out of 16 women, only four were legally available.

Let us now consider, diagrammatically, what could happen to fracture this predictable arrangement. We now consider two secure populations with intact clans where incest avoidance and the Westermarck and in-law effect is maintained.

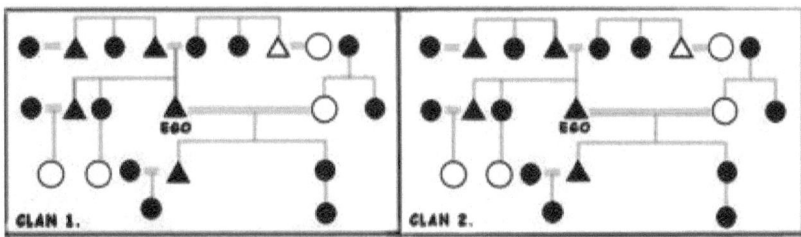

TWO CLANS, EACH ON A SEPARATE ISLAND

Figure 8. Separate islands

In the above illustration, each rectangle represents an island. Each island is inhabited by a small clan, called Clan 1 and Clan 2. 'Ego' in both clans may only father children by four out of 12 women in his clan. The only way that non-clan members can peacefully enter the other clan's area is via marriage or authorized immigration.

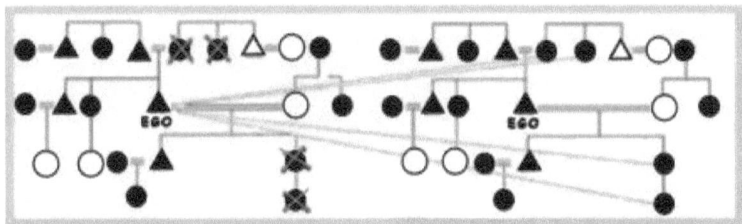

Figure 9. Violent invasion

In the diagram above there has been a violent invasion. Four women have died in Clan 1. All of them were off-limits for Ego in Clan1. Now four women from the unrelated clan 2 in the next island have entered territory no.1.

Suddenly 'Ego' has four more potential partners with possibility of legally fathering children by 8 women. His chances of fertility are therefore doubled, although the population numbers and density of Territory no.1 have not changed. What has changed is the fertility of the population in territory no 1.[200] (Actually it is more likely that males would invade and the women would find partners.)

What does a low-fertility system look like? Contrasting Central Australian Aboriginal or Korean with Leviticus system.

Figure 10, "Low fertility incest prohibition rules and affinal restrictions to 8th degree (Australian desert, traditional Korea)," is a graphic representation of the impact of an incest-avoidance system to the 8th degree. Represented by the white squares, the fertility opportunities are obviously very low. This system occurs among many peoples, including those in traditional Korea and the Central Australian desert.[201] In this situation, even with clans joined by marriage, "You" have only four possible partners among blood related kin and similarly, four possible ones among marriage related kin. The Australian film, *Ten Canoes*, directed by Rolf de Heer and Peter Djigirr, 2006, shows this situation, with a few married men living in huts with their wives (often more than one), and numerous unmarried men living in the unmarried men's quarters, and unmarried women in their separate quarters. Few of these unmarried men and women have much chance of marrying in the near future, if ever. The plot of the story involves the death of a married man, which allows a young man unexpectedly to marry the dead man's widow, whom he had coveted without much hope.

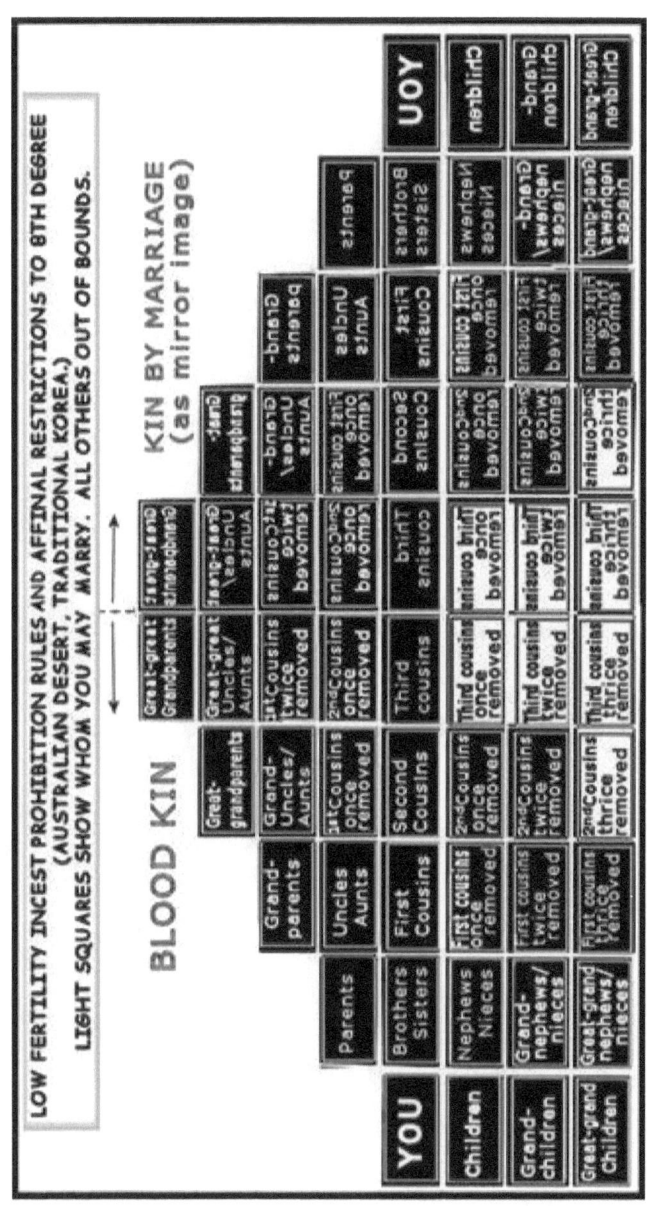

Figure 10. Low fertility incest rules

Figure 11: In Figure 11, "High fertility incest prohibition rules and affinal restrictions in ancient Hebrew society, according to Leviticus 18," the situation is almost the opposite of the Australian and Korean one. In Leviticus, out of 30 blood relatives, the rules permit 'You' to have 21 potential partnerships within your own blood kin, even allowing uncles, grand uncles, and great-grand uncles to marry nieces, grand nieces, and great grand nieces. Also, if your brother dies, you are expected to marry his widow. And those opportunities are expanded into the kin by marriage clan. This was a potentially very fertile system.

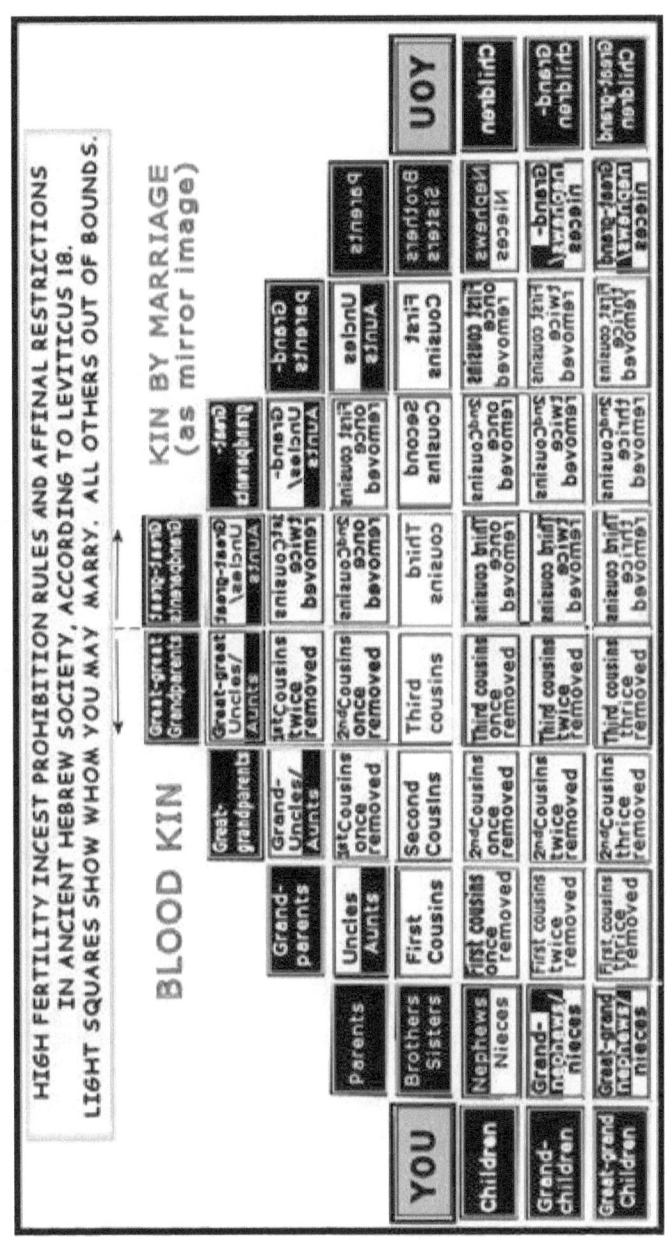

Figure 11. High Fertility incest rules

Patterns of incest avoidance and the Westermarck Effect in other animals and their fracturing may be extrapolated to human clan and village organisation, and then on to bigger societies where, for a variety of reasons, clan-based sequences have been interrupted and new layers have been built on top of fragmented old layers, around 'energy gradients' like river deltas. (There are many historical and actual examples, such as the Fertile Crescent, the Ganges Delta, the Nile Delta, Rome, Cairo, London, Sydney etc.)

In societies that do not rely on artificially modulated environments, population density reflects the fertility of the local environment. My hypothesis is that the environment affects hormonal feedback and hormonal feedback affects the fertility of individuals. The fertility of individuals is physiologically, socially and organisationally modulated by incest avoidance and the Westermarck effect, which is carried out more or less strictly depending on this hormonal feedback. Such patterns of partnering and reproduction are articulated in human cultures as laws and traditions. Human societies and their politics also reflect these patterns in their use of submission and dominance (as loyalty, obligation and power) in social hierarchies.

In 'modern' societies, characterized by national, regional, local and international population movement, and rich and poor food-sources imported from distant places, these biological population-spacing mechanisms have little or no relevant local biophysical and interpersonal political interface. That is, people don't think about them much and it occurs to almost no-one (has it ever been written about before?) that the opportunities for productive marriage in vast modern populations where a substantial proportion are constantly relocating, are amazingly numerous and varied compared to those in earlier societies.

Early peoples were born into territorial patterns which reflected the binary oppositions implicit in blood and marriage relationships, intersected by landscape concerns. All simple and traditional societies perpetuate these forms in their land-tenure and land-use planning. In the absence of horses and modern transport, fertility opportunities remain circumscribed.

In its most basic form, incest avoidance decrees that children move some distance away from their parents before forming their own households, since each human requires a certain amount of 'social' territory to mature and differentiate from his/her parent, and 'economic' territory to survive. Usually one or the other sex will tend to go and live nearer their mate/spouse's family, whilst the other sex will bring their spouse/mate back.[202] In a patriarchal society, it is usually the woman who goes to live with the husband's family. And, when a clan or more complex society (such as a tribe or village) reaches a certain population density (anecdotally between 300 and 500 people in a clan) whereby the territory it requires to sustain its inhabitants starts to reach beyond comfortable daily travelling distances, a new clan or other settlement will form.

If there is no-one within the available area who is not closely related, then some individuals will remain without inheritors.[203] Or, if there is not enough territory, in a subsistence culture, pregnancy will be avoided, aborted or any children produced will be neglected, destroyed or adopted out where possible, sold into slavery or indentured as low-paid workers and servants.

The inference is that, before complex agriculturally based societies developed, human families formed clan and tribal populations, in densities which reflected local biophysical opportunity. Yet another way of saying this is that the density of these populations and the distance between settlements was a function of the binary opposition of blood and in-law incest avoidance and the Westermarck effect,[204] interacting with soil fertility, climate, and natural features. These factors are tempered by the kind of economy the society had. In a hunter-gathering or herding economy more 'horizontal' territory will be required per person than in an economy that relies successfully on intensive farming, which 'mines' vertical territory in soil. In the case of coal and oil-based economies, extractive methods access not only vertical territory, but the goodness of soils, plants and animals, stored from other times, and may support many more people per ha than other systems, subject to fuel exhaustion and pollution.

In traditional societies land was inherited. There was no system whereby parts of clan, tribe or village territory could be aggregated and sold off to local villagers or to outsiders. Since land could only be passed on through inheritance, the pattern of settlement preserved its binary arrangement around blood and marriage. This was a biological algorithm that functioned within the local and regional biophysical environment, according to physical laws like the rules of thermodynamics as they mediate life-support.

Such was the foundation of a steady state society, where population varied little and there was little difference between the land rights and status of individuals. Numerous variations were possible within this 'topology',[205] for instance, it could be matriarchal or patriarchal, polygynous or polyandrous, but some things could break it. What can break this topology, changing systems within and outcomes, is the wider subject of this series, which ultimately will explore the evolution and impact of two different major land-use planning and inheritance systems – the British and the Napoleonic or Roman Continental European ones.

Culture and Structure

Humans are named according to their family lineage and their place of birth. Individuals may bear the name of their clan or tribe, which may also reflect the name of that tribe's territory, and special rights to hunt or obligations to avoid eating a particular animal. Such names are signposts to the blood or 'marital' relationships between the individuals of known clans and tribes in the area. Whereas names in a small stable society are like maps of local and regional genetic and ecological relationships, in many modern societies, names may have lost all intelligible relation to geography and clan. They are almost meaningless artifacts, except where they acquire new meaning through becoming attached to statistical data about individuals. Much of these data have arbitrary or specialised significance and are of little use to the individuals concerned.

Human cultures are quasi-transportable and this is one reason to explain why people transplanted to different new environments may behave in a counter-intuitive way. They are seeing their new environment through a cultural filter. The other qualities of dominance that express themselves in our family, clan, tribe and national politics tend to preserve established patterns. For instance, normal conformity to peer values and pressure makes it very difficult for most individuals to adapt independently to changes around them if the more dominant social forces seek to retain positions adapted to other times and climes. Peer pressure is strong cement that complements the power of dominant males and females over their more submissive fellows. Such bonds are integral to the order and structure of societies. The bonds involve power attributes such as territorial rights and obligations which motivate the incumbents to maintain position. Cultures that use money have to deal with vast networks of invisible bonds which have largely lost their original ties to geographically based land and resources. The laws of cultures – particularly those of land-tenure, marriage and inheritance - reflect these kinds of arrangements and are used to bolster them. Culture is our chief means of future planning.

The complex internal rigidity of cultures is yet another reason why it is difficult for complex societies, where both environment and culture are artificially mediated, to adapt their land-use planning and their population regulators to new conditions where flatter smaller societies might be in order. A case in point would be the need to plan for global climate change and petroleum depletion which, it seems, is better met by older, less mutilated cultures like those of Western Continental Europe than by the Anglophone settler societies and other products of colonialism with their broken land-tenure topologies.

Social Geometries

In a clan based society you would expect there to be a regular predictable geometric spatial relationship between persons, families and settlements, intersected by landscape and land fertility qualities. Habitations would remain with little change over generations, as they would be occupied by succeeding generations with few changes in numbers or needs.

Where clan series have been discombobulated, however, this geometric regularity would be broken. This breaking would then permit a new irregular mosaic pattern created through new incest avoidance and Westermarck effect patterns. Such disruptions are often detectable in villages and cities which have been subject to mass influxes, like settlement layers in archeological digs.

The mosaic would permit greater population density. Shanty towns and slums reflect this kind of demographic disorganisation in their high density and irregular agglutination. The practice of 'in-filling' - an Australian planning term to describe the re-zoning of built areas to higher densities - is another sign of demographic breaks in series.

Where original clan settlement patterns have been broken, memory has also been broken, so there would be incidental incestuous relationships due to new mosaic-pattern-dependent spatial proximity. Where memory has broken down, obligations, responsibilities and loyalties are also forgotten and new ones must be forged. Communities are vulnerable during these periods.

Spatial relationships may also be modifiable by custom-defined gender pathways. An example might be the aboriginal custom of closing your eyes in order to avoid seeing a close relative of the opposite sex;[206] gender practices involving the sexes spending much of their time working in different parts of the clan territory, or of one sex taking a very long route to do something which could be done via a much shorter route, but whereby the longer route keeps them away from the opposite sex; the practice of having separate houses and separate villages, separate land for the two sexes; the practice of purdah, where women have their own quarters and are contained and sheltered by an enveloping burka when outside that area (probably a reduction and symbolization of what was once female inherited land); and those modern gender divisions more familiar to Anglophone society, such as single sex schools and occupations that tend to be reserved for one sex or the other.

Some of us would also recognise that there has been a reduction in gender-specific customs in 'post-industrial' societies, with less gender specific entry into occupations and into institutions such as schools and churches. The 'uniforms' more familiar in 'mechanical' societies[207] where the range of head-dresses, scars and tattoos may be limited to one kind for each sex have been replaced by a host of different clothing styles, with elements of cross-dressing or of unisex apparel common.

The theory I have developed about incest avoidance and population dispersal does seem to have strong fractal elements. Australian environmentalist, Greg Wood, was reminded by my theorizing of the laws of thermodynamics and their impact on order. Remarking that, although most energy expenditure leads to disorder, life restores some order before it degenerates as well, he suggested that my theory of binary oppositions as a limiter of ultimate clan, tribe or city population size, might also apply to individual organs and to the phenomenon of cancer – whereby organ tissue multiplies and entropies (becomes disorderly).

The notion of cellular division and then sexual differentiation are like fractals on a continuum, with cellular division in protozoa at one end and sexual dimorphism and incest and Westermarck avoidance, and separation between the discrete populations represented by clans, tribes, cities and nations at the other.

New Theory of Demographic Adjustment

I am aware of no other sociological theories which compare the effects on population size of different systems of land-use planning, transport, marriage and inheritance. This is despite the fact that my theories derive from observations which were the building blocks of much anthropology prior to the 1980s. The material I use to identify algorithms of population spacing in other species is recent and was developed for other purposes, mostly to establish whether incest avoidance occurs in non-human species (a very rich area), and then to measure its effect on genetic health.[208]

Perhaps due to the assumption that genetically programmed incest avoidance has evolved as a norm because it diminishes the risk of deformities and other inherited problems, the effect that avoiding incest has on territorial division and population spacing – i.e. on population growth and its spatial distribution – seems so far to have gone unremarked in these research conclusions. We read that incest avoidance is achieved by dispersal but this is not turned round to observe that incest avoidance *causes* dispersal.

In my theory, if algorithms of population spacing occur in other species they must be biologically and genetically based. I then hypothesize that incest avoidance norms in our own species almost certainly arise from similar autonomic sources. As I have suggested, in studies of other species, there is good evidence that these algorithms adjust to hormonal responses to sensory and alimentary feedback from the environment where availability of territory may be measured by distance from close relatives. Potentially fertile offspring who are dominated by close relatives cannot find territory of their own and non-incestuous, non-Westermarck effect opportunities for mating are not taken up. There is evidence that hormonal suppression also occurs in similar circumstances in humans. For instance it has been observed that girls raised in the presence of stepfathers menstruate earlier than those raised in the presence of their blood fathers.[209] Incest is more likely to occur in such cases as well, in ours and other species.[210]

In human societies, indications of incest avoidance as a major factor in population spacing are most obvious in rules for marriage and inheritance laws where marriage is defined as a union which gives rise to legitimate children who have family rights to inheritance and, in modern societies, citizens' rights to land or an equivalent, such as employment or unemployment benefits, or public housing, and the vote and a passport. In a subsistence society a legitimate child is born with land rights; illegitimacy equates to being born in a situation that does not entitle the person to land. Not having land-rights means that the person has no means to survive independently. In Book 2 of this series, on Land Tenure and the Origins of Capitalism, we see that societies have evolved where the land-tenure system allows legitimate children to be disinherited. (The great Anglophone societies, starting with the UK, are products of this system, which they have transmitted to many colonies. It is the basis of private property and of capitalism.) In such cases the State may accord them certain rights of citizenship, often of a marginal variety. Industrial societies *rely* on a working class which lacks sufficient land-rights for independent self-support.

These differences in social structure and function are well illustrated below by the authors of an introductory unit for the course, "Aboriginal cultures and the land".

"[...] because Aboriginal societies were highly egalitarian, there was/is never any need to produce 'surpluses' so that the labour of many could/can sustain the wealth of a few—a primary characteristic of what we usually call 'civilisation', where oppression of both our fellow humans and of the natural world are fundamental to what passes for 'civilised' society. [Add slavery, widespread conquest by warfare, writing, and building in stone, and you get a 'Great Civilisation', such as ancient Rome or China!]"[211]

The same introductory text affords us a good compare and contrast between a society structured primarily by kinship and one commercially structured. The theoretically almost unlimited choice of partners in 'Western' societies, compared to the very limited choice in aboriginal cultures, is also remarked upon:

"Kinship in Western society tends to be fairly narrowly defined. The isolated nuclear family is of great significance; other relatives may also be important in a secondary way, but in general this also involves a restricted range of people.

Many aspects of social life are organised through other institutions—the economy, the polity, the education system, etc.; kinship is limited to a particular sphere of social life (the family), rather than being a pervasive structuring principle as it is in many societies.

Western societies emphasise patrilineal descent. Despite some recent changes, names are allocated patrilineally. In the past there was also a male bias in the inheritance of property. [Note that this was not the invariable case throughout European cultures – SMN.]In some cases there may still be traces of this when sons rather than daughters inherit the family trade, farm or business.

'Coming from a good family' can still be important, but this is not highly structured in terms of particular, defined lineages (it was much more important in the past, when the monarchy and the aristocracy were of greater significance).

Cross cousins and parallel cousins are not distinguished. Residence after marriage is neolocal. Exogamy only applies to a small range of close kin (those to whom the incest tabu applies). Apart from this the choice of marriage partners is in theory unlimited—there is no formal system of endogamy. Informally, however, people do tend to marry those of their own ethnic and socioeconomic status.

Generally, kinship and descent tend to be overshadowed by other factors—formal education, the bureaucratic organisation of work, job mobility, social security, institutionalised health care systems, and so on. Apart from the nuclear family, groups based on kinship or descent do not play a large part in the organisation of [Western] society."[212]

All human land-use planning, marriage and inheritance laws explicitly incorporate incest avoidance to some degree or other, including explicit exceptions.[213] The permissible variations on incest avoidance have been iconicised in religion and codified in law in every society.

Institutionalised exceptions do occur and are discussed below.

Exceptions also occur accidentally with isolated populations. The isolation of those populations and the built-in overshoot factor mean that these exceptions are very unlikely to normalise in wider populations. We will discuss theories for the very famous example of Easter Island, under the heading of isolated populations in the Chapter, ## "Pacific Islander and Modern Land Tenure and Inheritance." Our conclusions are almost certain to surprise and hopefully to satisfy.

Institutionalised Exceptions to Incest avoidance affecting land-tenure

Where exceptions to incest avoidance laws are institutionalised in non-isolated populations, they always seem to be related to securing land and other power bases.

We have already mentioned the Bible as a source of commonly prescribed marriage rules. A widespread example in some societies is the 'levirate', in which tradition expects a man to marry his deceased brother's widow if they live on the same estate and the deceased has no male heir. This keeps the land in the same family lineage and any new descendents may be treated as children of the deceased.[214]

Another form also prescribed in the Bible was to marry daughters to their father's brother's sons (the girl's first cousins). This was also a way of preserving land in the same male line because the male cousins descend from the same grandfather. This is known as 'parallel cousin marriage and lineage endogamy'.[215]

Even slaves (presumably as family possessions) would sometimes be co-opted as legal spouses for the purpose of securing an heir.

Anthropologist Brian Schwimmer,[216] provides a useful visualisation of these arrangements, which I have adapted[217] below.

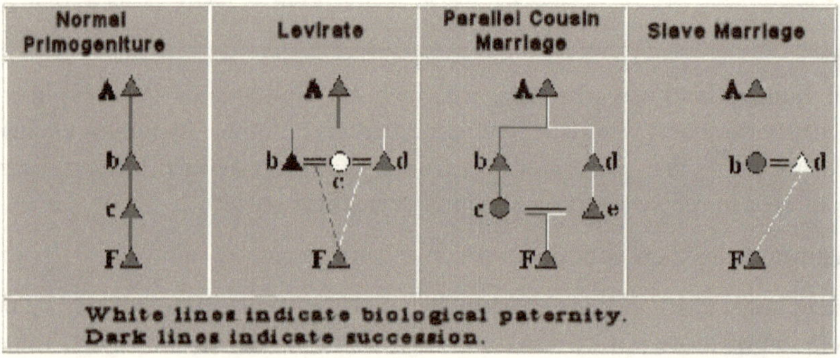

Normal Primogeniture	Levirate	Parallel Cousin Marriage	Slave Marriage

White lines indicate biological paternity.
Dark lines indicate succession.

Figure 12. Parallel cousin marriage and Lineage endogamy with slave marriage. [218]

The case of Egyptian pharaohs where brothers and sisters were married is well known. Less well known is the practice of marrying brother to sister among Ancient Greek colonials in Africa. Gender streaming, where the Ancient Greeks tended to rear male and female children in separate living areas, probably removed the Westermarck effect impediment to this practice. First degree incest may also occur through the union of closely related cousins in caste intermarriage, as practised among royalty. Caste systems also evolve in colonies, where the colonisers preserve land-acquisitions, power and solidarity, by excluding the colonised as marriage partners. We will see this in Book 2 on the origins of capitalism, in the behaviour of the Norman class in England from 1066, where caste separateness was the structural underpinning of the feudal system of land-aggregation and the social structure of work.

Domestic animals in modern human societies

Exceptions to the incest and Westermarck effect are also observable in the behaviour of domestic animals that live in human societies. These animals tend to practice instinctive incest avoidance when raised normally in the wild. For instance, the normal familial society of wolves contains one breeding couple closely associated with non-breeding related males and females in a close-knit extended family.[219] But, among the descendents of wolves – domestic dogs – the clan arrangements are rarely preserved and dogs often don't know who their fathers, sisters, cousins and uncles are, due to the dispersal of litters. Such animals have more opportunities to mate because the order of the clan is absent. There is similar social fragmentation and loss of kinship knowledge in countries which are 'melting pots' or war-torn, where people are subject to frequent geographical and social dislocation. The children of artificial insemination, be they human or cattle, have no chance of consciously avoiding blood-relation incest since they don't know their fathers, siblings or uncles and aunts.

Runaway populations of felines, canines and humans may have similar causes. In the 'settler societies' of the United States, Australia, New Zealand, and Canada, indigenous populations have been greatly disorganised. So, however have the settler populations, perhaps even more so. The result is a chaotic soup where many do not have any relationship with the land or the incumbents with which to orientate their incest avoidance gyroscope. Furthermore, territory is defined by contract and acquired through money, not through birth-right. What effect must this have on the innate skill we infer here to be normally available to members of populations in stable situations to gauge safe levels of reproduction, well within carrying capacity and comfortable density?

Some other questions arising

Kinship theory has demonstrated that humans and other animals are more careful of the feelings and rights of their blood relatives and clans. I think that we have to consider the application of the Westermarck Effect here and consider that it might be better to restate this to say that humans and animals are more careful of the feelings and rights of those they grew up with. If the ties to these people don't exist, or new Westermarck ties are not mediated by a traditional biological framework of dominance, what effect might it have on justice and equity overall? Even if we have a sociable concern for our neighbours, what is the effect of competition when the loss of our kinship-territory gyroscopes causes situations where we are forced to compete for scarce land, resources such as water, and services – such as higher education, hospital beds, and well-paid employment? It was this kind of disruption of kinship organisation and loyalties that permitted the formation of feudal system and industrial society with its stark social division of work and property ownership. (See Book 2 of this series on Land Tenure and the Origins of Capitalism.)

These risky situations are not confined to the 'developed' Anglophone societies; they appear to be at the roots of the chaos in many other societies, notably those affected by the British system, for instance, in those Pacific, African, Indian countries, states and regions, where dispossession has been profound, sustained and repeated. The situations in 'developing' countries which have managed to preserve kinship series and their relationship to land is usually better. China is somewhat better off than India, albeit China suffered profound dislocation by Mao and subsequent post revolution regimes; Japan, which might have been overwhelmed by war and colonisation, has retained many kinship series and the related aspects of land-tenure sufficient to preserve its forests and to be among very few nations with islands in the Pacific that do not suffer from overpopulation; its citizens have achieved higher standards of living than most Europeans. Pacific islands in French territory also seem to suffer from less overpopulation.

The relationship with the land is our normal means of territorial feedback regarding available necessary space from which to derive a reasonable living. In the Anglophone first world countries where populations are growing rapidly, this normal relationship has been replaced by abstract environmental feedback, via media, opinion, arbitrary statistics and political marketing. As suggested by research cited earlier in this chapter,[220] underpinning all environmental feedback is probably reliability of nourishment. International trade has obscured seasonal variations in available foods and abundant fossil fuel has removed seasonal and regional limitations to available calories. One result of this is a much greater intake of calories, particularly of rich fats, in some first world populations. An impact of this increased fat intake has been obesity and decreased age of menarche. Another impact, also apparently acting on hormonal feedback, has been a rise in diabetes. This implies a fall in life expectancy. Who knows what the ultimate equation might be in response to the chemical soup of the massive populations of the 21st century?

How disruption of incest avoidance system would deliver more fertility opportunities

Effect of transport systems

A belated realization in writing this series[221] was the multiplicatory effect of new forms of transport on fertility opportunities and size of settlements.

For a very long time, of course, clans and tribes had only their feet to carry them in their search for space and mates.

The introduction of boats and the taming of riding horses, pack animals and, with the wheel, cattle to tow loads, must have vastly increased both fertility opportunity and the ability to supply larger settlements, along with new storage and food preservation methods. We will see how the development of ships and sea routes when overland spice routes were closed to Europeans, coincided with the rise of a new international slave and bullion trade that set Europe on a new course of imperialism, known as the Trade Wars.[222]

The technology to smelt iron with coal was fundamental to the railway. With the development of rail in industrial and colonial societies, it was possible for people to supply, settle and interconnect land at much greater distances. This permitted the formation of many new households along rail routes, creating a certain predictable geometry.

The post World War 2 "baby boom" was enabled by the popularization of the automobile in a petroleum economy which, mass-produced, made new territorial conquest available to many more individuals, along with new fertility opportunities.

At the same time aeroplanes with ever-decreasing passenger fares, connected the populations of continents en-masse.

Failing effective land-use and inheritance system reforms to stop runaway population growth, the thing most likely to stop this is petroleum depletion, which could reduce the ability to service and connect 20th century style settlements back to before the development of most railways, perhaps to 18th century levels.[223]

War and colonisation

How would the disruption of the incest avoidance system, through war, colonization, and natural disaster, work to deliver more fertility opportunities? Consider that, where populations of any animal are destroyed and families and clans dispersed, parents become separated from children and sibling from sibling. Sometimes entire peoples perish. If there are survivors though, under such confused conditions, they grow up without knowing their parents, their siblings or their ancestral lands.

Land-rights are left undefended and now people from other places may move in and use the land differently.

Disruption could be so drastic that, if the opportunity arises to procreate with an unrecognised parent or sibling, or 'in-law', no incest avoidance or Westermarck response, which in humans requires close association during the first 30 months of life, would have been initiated.[224] The spatially organizing parental and other blood relative dominance would be missing.

Fertility opportunities may thus be dramatically increased after an initial die-off since individuals may be thrown together in groupings and in new places where every individual is a stranger and therefore all those of the opposite sex could be legitimate potential mates. The stress for the displaced and harassed to organize through bonds with others of their kind would be a natural survival strategy. The result could be a population explosion – in humans, kangaroos or any species which can normally be shown to practice incest avoidance. This could explain the extremely high rates of fertility in troubled African states, for instance.

This concern may seem exaggerated to the modern reader if they come from one of the many structurally disorganized modern societies, where concern for kinship has a low priority in the economic order that rules those societies. This is particularly true of the Anglophone settler societies. In such countries clan organization has been obscured, fragmented and subsumed to major and minor transmigrations which have become an integral part of a finance and property 'growth lobby'[225] which seeks to perpetuate them.

People locate away from their home-towns to be close to employment and form routines and attachments that bind them to new places and organizational structures (such as industrial workplaces) and alienate them from their roots. They marry people who come from completely different backgrounds, with different values and loyalties. Christmases are famously times of existential angst where sons and daughters go 'home' to places and people whose lives are now completely separate from theirs. International and inter-regional migration is facilitated by cars and planes and dictated by economies where employers reserve the right to demand that spouse relocate with corporation. Spouses divorce if their loyalties to or dependencies on different employers over-ride their commitment to marriage. Ownership of land and ties to place are a hindrance to mobility in such cases as are the related ties of marriage. Renting, divorce and serial families become normative. We will see later how this situation benefits modern capitalism, not only via transaction, conveyancing and service fees from lawyers. The measure of the GDP has often been criticized as a poor measure of wealth because it measures any activity, good or bad.

What it also shows is that many activities that create profit will persist whether they are good or bad, even if they cause immense suffering.

In a society where there is constant relocation within a huge population in a large territory, the probability of accidental incest seems so low that the risk arouses little or no concern. Rituals of avoidance beyond the primary family in a confined household are quasi-nonexistant. If there were concern it would be for the possibility of producing genetically compromised children. Modern society provides a technological solution for this though. Where individuals from an identifiable or self-identifying ethnic group or who believe they are cousins contemplate marriage, blood tests may be carried out to see if either is a carrier of an inheritable disease with high prevalence in that group – such as sickle cell anaemia in descendents of Mediterranean peoples. The potential parents may then be counseled not to have children, or to have amniocentesis and consider aborting any affected foetuses.

As I have stated, the concern I am flagging here though is not of inherited diseases, but about unintentionally or unwisely multiplying the rate of population growth beyond safe levels, in terms of equity and environment. In fact unstoppable population growth and overshoot of resources, such as water, along with homelessness, are already fact in those huge modern first-world Anglophone societies that are constantly in structural flux.

This seems to be what happens in all disrupted populations - from kangaroos to cats, from mice to men.

This chapter was about a new theory of population which is based on genetically based algorithms and food supply. In the rest of this book and the other books in the series, this new theory will be employed to re-examine history with particular reference to theories of societal collapse and the development of modern expansive and conservative economies and populations.

CHAPTER 5: PACIFIC ISLANDER AND MODERN LAND TENURE AND INHERITANCE SYSTEMS.

Flat, steady state societies, deriving most of their energy directly from solar, wind, water, soil and local plants and animals are the norm and they may last for a considerable time. For complex societies, history really consists of booms and busts, not some continuous upwards curve of economic growth or progressive enlightenment. Tribes aggregating around massive energy convergences, on deltas, for instance, where there is plenty of soil for agriculture, wood for fuel, and animals and plants to eat may come to rely on growth, rise to great heights and then fall. If totally isolated, they may collapse into mysterious remnants. What keeps some from totally crashing is their proximity to neighbours, who take over what is left and animate it, as has happened over centuries in Europe.[226]

On the back of massive fossil fuel stores of coal and oil, the effect of new forms of transport has enabled increasingly rapid population movement and the establishment of huge 'settler states' like the United States, Britain, Australia, Canada, as well as Africa, Central and South America, Indonesia and other colonial creations where mass immigration tends to be termed refugee and asylum-seeker movement. Each form of transport that has permitted greater numbers of people to move longer distances has multiplied population growth.

Fuel supply, transport and huge populations

History, since the commercialization of coal and petrol as industrial fuels, records population growth, expansion and consumption at previously undreamed of levels. At the moment, due to the world's fuel-tank being still around half full of petroleum, which has permitted total commercial connection, this is the situation on a global level. No country or region is permitted to sink into complete collapse because there is always some industry somewhere that will take over whatever is still viable. The half-full fuel tank cannot continue for much longer, however, as population and activity increase. When the global supply of cheap and abundant fuel fails, this time it is feared that collapse of complex society will occur for the first time on a global level.

Some countries have defenses against overpopulation

Not every large economy is on the same trajectory of population growth and expansion, however.

Two different major first world social systems on this planet exist: one which dooms people to saw off the branch they are sitting on and the other which keeps the branch we are all seated on intact.

The second system has retained properties of an earlier land-tenure system that is crucial for limiting growth in line with ecological circumstances. The first system has lost these crucial attributes and is, in effect, broken.

This ancient land-tenure system by which human societies can adapt to limits is found wherever societies have existed for hundreds of years in a steady state. It is to be found among hunter-gatherer tribes, horticultural communities, herders and nomads in every part of the world that has not been transformed into a growth system. Although this system is not confined to the islands of the Pacific, it has been very well described there, and the many different islands have formed a kind of laboratory of variations on social rules over time and space. The Pacific Islands were colonized a century or more after Africa and India.

It is easier to see how things work on a small scale in an isolated space and the rules of steady state societies first became obvious to me by reading about traditional land-tenure in Pacific Islander societies. After that it became easy to see how it had applied in larger areas, like Australia, Malaysia, and pre-colonial India, North and South America and Africa, where different clans, tribes and villages connected and interacted with each other whilst remaining intact and distinct. Because I first learned of it in traditional Pacific Island examples, I think of it as the Pacific Island Land Tenure and Inheritance System. It could more generally be referred to as the 'steady-state topology'.

The Steady-state Topology or the Pacific Islander 'rule'

Nowhere is the ability of a system to limit growth more important than on isolated islands. The Pacific Islands were the last in the world to be severely disorganised by colonisation. The destruction of their original societies was documented by many sources, including religious, commercial and anthropological. Australia is to be counted among the once isolated islands of the Pacific and Australian Aborigines used the same system as all the other islanders. Joseph Birdsell (previously cited) describes an example of this in Australia.[227]

'Modern' Australia, however, is a product of British colonial history and has been totally made-over according to the British capitalist land-tenure and inheritance system. For all that it is bigger and more resourced than its smaller neighbours and for all that the white population far outnumbers the original black and brown Aboriginal populations,[228] Australia is dogged by the same exploitation, overpopulation, and ruin as the other Pacific islands of our time. Because it is bigger – although mostly hot desert and rangeland - and has access to international reserves of fossil fuels, its ruin takes a little longer, although the same access to fuels and mechanisation also speeds up the process.

Isolated Populations: Easter Island as an exception to the Pacific Islander norm

Incest avoidance laws are crucial components of steady-state Pacific Islander land-tenure systems, which occur in relatively isolated, small populations. The success of the Pacific Islander land-tenure systems reflects natural selection. Societies which did not practice incest avoidance would logically overshoot and then collapse. Their isolation would select such socio-biological systems out and would preference the incest-avoidance and Westermarck Effect genetically transmitted tendencies into the larger human population.

To illustrate this, let us examine the legendary collapse of Easter Island society on the island now called Rapanui. The Easter Island collapse hypothesis is of the late colonisation by a single party of Polynesians of one of the most isolated places on earth,[229] in the Pacific more than 3,600 kilometers from the nearest population centers, in Chili and Tahiti.[230]

This fabled collapse allegedly and conveniently occurred just prior to the European incursion, and so provided fertile ground for hypotheses. Here is one more, from me.

With or without Jared Diamond's collation of biophysical characteristics of islands associated with more or less deforestation, hence collapse,[231] or theories about rat overpopulation,[232] the collapse of Easter Island could be explained by a variation of incest avoidance in a population formed from fewer than one hundred – possibly only around 40 people with little genetic diversity – in total isolation. This was the situation at the beginning of the lost civilisation of Rapanui, according to legend. If that was really the case, the odds were vastly ranged against their society surviving for more than a few generations at most, unless incest avoidance and the Westermarck effect were abandoned. If they were abandoned, however, barring catastrophic events (comets, plagues and volcanic annihilation, for instance) the mathematical outcome would be predictable – a population explosion.

The story goes that the first settlers, arrived at the 117 square kilometres island in one single flotilla containing between 40 and 100 passengers or, at the very outside, up to 200, in approximately 390 or 400 AD.

These small numbers find some support in mitochondrial DNA examination of a reasonable sample of human fossil remains which, according to one source, revealed only one female lineage in 2006.[233]

The civilisation, although tiny, had writing and monuments. According to the pre-European collapse theories, it lasted less than 900 years but longer than many civilisations we currently consider important. Legend has it that the leader of the expedition divided the land up between his sons. By 800 AD the forest on the island was already severely depleted. At peak monument production stage, between 1000 AD and before 1500 AD, theoretical reconstruction has the population numbers at somewhere around 7000.

In about 1500 AD the population is believed to have risen to its peak of around 10,000, crashing soon after.

To reach 7000 and then 10000 in population could only have happened, assuming the islanders were closely related, if they had overcome the ordinary Pacific islander land-use planning and inheritance traditions outlawing marriage between close relatives and different castes. In fact, just stopping the original population from dwindling would have required quite dedicated pursuit of incest plus a lot of luck.

Apart from the math, there are documents suggesting that traditions may have evolved to encourage marriage between close relatives. A caste system on the island also would have encouraged endogamy.

> "With an initial population base of perhaps 200 persons, some kind of in-breeding was inevitable — especially among the royal clans. While it is known in the rest of Polynesia that deformed or otherwise "abnormal" babies were killed immediately after birth, there is no historical or legendary evidence for this practice on Easter Island. This pertains mostly to physical deformity, and since mental retardation resulting from in-breeding can take years to become apparent, it's quite possible that such offspring would have survived at least for a time.

"There is some evidence of endogamy (marrying within one's group) among certain island clans. In order to preserve blood purity, it is said that members of the Urumanu tribe were not allowed to marry members of the Miru clan (and vice versa). Based on medical research done by the Canadian "Medical Expedition to Easter Island" in 1964, there is a statistically high percentage of Easter Islanders with six toes on each foot — and a peculiar degeneration in the knee joint. (Some members of the Miru were said to "walk funny" and were given the nick-name *ngapau*, meaning "bow-legged".) There is osteological evidence of in-breeding (at Anakena, the legendary landing site of Hotu Matu'a and the home of the royal Miru clan) — and early Easter Islanders generally had the most prevalent dental caries of any prehistoric people." [234] [High rates of dental caries have been linked to endogamy.[235]]

Some other practices noted point to a society that adapted its mores to the initially very difficult task of boosting population from a very small base. In the quote below, child marriage was allegedly one such practice. Child marriages are often concluded to cement political relationships, but early marriage also increases fertility opportunities. The practice of lending or leasing one's wife is also described in the quote below, and linked to a general desire for offspring.

"The marriage ceremony is performed by the acting priest in the church, but the practice is permitted with children who have not reached the age of puberty, and the betrothal is conducted by parents, the relations of the female paying a stipulated amount, generally in food to be consumed by the friends at the feast given to celebrate the event. It is not certain that polygamy ever existed, but an ancient custom permitted the husband to sell or lease his wife to another for a stated term. On account of the disproportion in the number of the sexes, celibacy was a matter of necessity, and probably originated this custom. Love of family is a strong trait in their character; children are fondly cared for, and the desire for offspring is general."[236]

How did they overcome the drag on sexual attraction that comes with the Westermarck Effect, which must have been rife on Easter Island? One would have to assume traditions to keep male and female children from mixing during their early lives. To expand the damping of the Westermarck effect throughout the immediate family, traditions keeping both sexes apart as much as possible would have assisted.

Although no-one is sure exactly why the population crashed after reaching around 10,000, according to these more or less hypothetical conditions, one assumes that, having survived at all, overpopulation became the compelling destiny, due to the loss – through absence of choice of marriage partners – of the brakes that dictate space between settlements by responding to biophysical constraints.

By 1700 AD many sources believe that the population dropped to about one tenth or one quarter of its former number – around 2,200 or 1000 people then. The first reported European contact was in 1722.[237]

The giant Moai statues

One can even hypothesise a related role for the apparently bizarre importance of the giant statue-sculpting which some commentators believe reach frenzied heights right at the very end, leaving many partly carved monoliths in the quarry, including one which would have been over 60 feet tall.

There is a particularly poignant arrangement of statues seen from a certain point of view off the coast of Easter Island that reminds me of SETI – the 20th C project that searched for life in outer space. In this poignant arrangement on the island, the giant statues line the shore, as if they had been placed there in a desperate attempt to make the island's inhabited state visible from a great distance. A purpose could have been to attract passing seafaring islanders to its shores.

Were the Easter Islanders keen to lure victims to their cannibal lair? This seems an unlikely objective of such a long-term and wide-spread monument-building tradition (present in different forms throughout the Pacific). It seems probable that they were keen on doing trade and hitching rides to other places and in advertising for immigrants as partners for their sons and daughters.

Easter Island was so isolated that it is thought possible that no further traditional Polynesian colonisers ever arrived. Despite these inauspicious circumstances, this one party of Polynesians purportedly seeded a complex civilisation which reached numbers between 7000 and 10000, profiting from the abundant sea-life and productive vegetation. But the downfall of the Easter Islanders apparently lay in their very success, for they reputedly ate themselves out of house and home, finally dwindling to a handful of incestuous cannibals, surrounded by giant statues which had been produced by their dead civilisation. Having chopped down all the trees, they could no longer make canoes to go fishing. By 1700 AD Easter Island's population had purportedly dropped from an estimated 10,000 to around one or two thousand. This reconstruction hangs largely on the belief that no visitors landed until the first Europeans, whose arrival is formally dated 1722, some of whom described the straggling survivors of the former nation.

This form of death to a society by over-consumption has been held up as the likely or even as the inescapable fate of modern man and consumer society by writers like Jared Diamond[238] and Tim Flannery.[239] Tim Flannery called it "Future Eating". In a series of international petroleum depletion discussion e-lists led initially by Jay Hansen of www.dieoff.org, Easter Island has been held up to illustrate the proposition that humans will *always* overshoot their environmental carrying capacity.

One wonders how many small island populations suffered similar fates to that of Rapanui and left even fewer traces of their passage? And how many simply allowed themselves to die-out without reproducing?

But then again, so many island societies survived, not just for hundreds, but for thousands of years. Ironically, the fate of this allegedly spectacularly unsustainable Pacific Island society highlights the spectacular success of so many others, and the evolution of a default steady state system.

Something wrong with this picture

But wait – there is something wrong with the whole picture. Rapanui is supposed to have lasted about 900 years before hitting the wall and collapsing, just before Europeans arrived. 900 years may be a short time compared to the 40,000 of Australian Aborigines and the 3,000 of Hawaii and various island civilisations, but it is still a very long time compared to the United States and modern Australia. For an incestuous build-up of population to occur so slowly, there would have to be centuries of false starts and then a sudden crescendo. Or there would have to be a slow build-up and then a catastrophe.

In fact those documents that I cited supporting my hypothesis of the normalisation of incest are probably quite unreliable, even though some of them purport to come from a Medical Expedition to the island.

In "Inbreeding and Surnames: A Projection Into Easter Island's Past," Antonio Gonzalez-Martın, Clara Garcıa-Moro, Miguel Hernandez, and Pedro Moral, claim that they were able to show that the Easter Islanders used strategies to avoided incest which were carried through by their descendents in the 20th century.[240] Of course the islanders could not help having a relatively small common gene-pool, but in the twentieth century, when their numbers had been drastically reduced by slaving and disease, they were apparently still using strategies based on geographical tribe or clan locations and names to avoid marriage between relatives between second cousins and nieces and uncles etc., which is a higher degree of incest avoidance than that prescribed by law in most modern societies and in the catholic church.

"Other ethnographic information on the behaviour of the inhabitants of Easter Island toward consanguinity is provided by Englert (1954), who stated that kinship grades were of father lineage, determined by distances between generations. This is the same as the Roman and civil Chilean methods. The only marriages allowed were between third cousins, a rule more severe than that of the 1917 Catholic canon. Marriages within closer kinship were regarded as incest. Finally, studies based on genealogical reconstruction going back to 19th century families certify intertribal exogamy proposals"[241]

Although the authors admit that the islanders must have practised incest in order to survive initially, they are stressing that incest avoidance reasserted as the norm[242] as soon as survival became possible without incest.

"The demographic history of Easter Island reveals a significant population oscillation throughout its history. With the population at its lowest, consanguineous relationships must have been inevitable. The probability of the occurrence of consanguineous interbreeding during the colonization of the island, as well as during other demographic processes as the population decreased, must also have been high. Above all, these circumstances must be considered exceptional; under normal conditions, the island's population would have been large enough to develop strategies with which to avoid consanguinity."

How could the islanders have avoided incest and have bred up to 10,000 people from 40 or 100 original settlers? Either they didn't practice incest or their numbers never attained the peaks the legend says they did. Indeed. There really is no evidence that the Easter Islanders ever hit 10,000 or 7,000 or even 5,000.

"Easter Island presents a problem because the case for demographic decline caused by anthropogenic environmental devastation is insufficiently documented at critical points. [...] All estimates of the peak size of the prehistoric population are entirely speculative; it may never have exceeded the 2000-3000 that can be estimated from early historical records."[243]

And, there is another suspicious thing – that the collapse is supposed to have happened just before Europeans arrived, when we know that the arrival of Europeans coincided – and in fact caused – the near collapse of so many other Pacific island societies and populations. The European Trade Wars over South America and other colonies had been going on for more than a century. Hard to believe that no-one had stopped off at Easter Island before 1722.

What if the basic facts are wrong about Easter Island's collapse?

The dates of the events and the sources of historical observations are mostly so uncertain that, if we were to use Occam's razor rule, and select the most likely explanation, then we would suspect that the society of Easter Island was destroyed by European contact, notably slavery – which was the major European industry of the time. We would suspect that the dates of first contact were wrong. We would suspect this because there was plenty of traffic around this area anterior to its official discovery.[244] There are also many documents about European slaving on the Island and because we know, if we seek, that this is what has happened to most Pacific Islander civilisations and societies.

Benny Peiser has written an impressive article deconstructing the popular Easter Island collapse hypothesis.[245] He finds that there is no compelling archaeological evidence of societal breakdown prior to the 18th century, which was the time of the influx of European ships. Fifty-three known ship visits occurred between 1722 and the beginning of the Peruvian slave raids of 1862. How many unknown ships, during this high period of the Trade Wars and the slave trade, where France, Britain and Europe vied with Spain for South American and other trade rights, passed that way is anybody's guess.

Peiser also lines up the evidence for time and degree of tree loss and finds it vague and conflicting.[246] It is even possible that some palm trees remained in the 19th century.[247] Toromiro [shrubs or small trees] remained until the 20th century. They provided wood for housing, small canoes and wooden carved figurines. Why couldn't they have been used as rollers and levers to move the statues as well?[248]

He also finds fuel for an argument that the stone statue production, rather than being abandoned just before Europeans first officially visited in 1722, continued, with statues being hewn right up to the time that most of the Easter Islander people were dragged away as slaves in the last quarter of the 19th century.[249] In fact this information was once well-known, according to Peiser, who attributes its obscuring to Thor Heyerdahl's writing and the popularisation of his theories, which he sees as ideologically driven.

He has some strong questions about Jared Diamond's forest depletion dates. He argues that the sources Diamond used put the loss of palms at around 1320 AD but that Diamond stretches this by about 330 years, basing his hypothesis on situating peak forest-destruction in 1400, yet having the Easter Islanders only 'reduced to burning herbs, grasses and sugarcane scarps for fuel" so much later, in 1650.[250] Peiser criticises Diamond for suggesting that lack of large timber and rope led to the end of the stone statue cult, but failing to account for the persistence of the toromino tree on Easter Island. According to Peiser, JJ Palmer of 1870 wrote that he saw the remains of large boles from palm trees as did a co-traveller, Dundas.[251] Fleley also wonders if the palm persisted until sheep and goats finished it off in the 19th and 20th centuries. I wonder if the ships didn't cut down the timber and haul it away. The western world was always on the look-out for timber, as we will learn from looking at Roman and British history in book two of this *Demography, Territory and Law* series.

Several accounts by 18th century visitors to the island describe richly cultivated land and healthy well-fleshed people. Peiser says the island's fish stocks are very rich – but he is writing that a long time after the hypothesised crash and they might have recovered, hypothetically. The Canadian Medical Expedition of 1964 noted that the situation of Easter Island was out of the way for migratory species both from cold and tropical areas, but nonetheless caught over 2,500 fish specimens over the two months they were there.[252]

Cannibalism tales questioned

Peiser finds no archaeological evidence of starvation or cannibalism at Easter Island and that there is no evidence of any overpopulation at the critical times.[253] The population, in his opinion, may never have exceeded 2-3000 people before the Europeans came. He argues that there was a tradition of attributing the statues to some other ancient people – the "Long Ears" – who had been displaced by another tribe. He believes this misattribution could have been motivated by a belief that the Easter Islanders were an inferior race and interprets Heyerdale's theory that another people settled Easter Island prior to the current islander-race was rooted in racism. He also describes Captain Cook apparently having difficulty believing in the ability of the Polynesians to create the statues on Easter Island.

Peiser suggests that the few islander-survivors of slavery and colonialism who were taken over by missionaries in the 1864s gave the missionaries the history the missionaries wanted or expected, which was one that exonerated the missionaries' European lay colleagues and themselves. It seems likely, from his argument, that the 'Long Ears' were the Easter Islanders themselves and that the tribe that wiped their culture out was the European tribe. He also thinks that the tales of cannibalism could have been made up by the Easter Islanders as a defensive measure, to give the missionaries the impression that they might be more capable of fighting off any renewal of European excesses than they really were. In 1868 the islanders gave in and converted to Christianity within 4 years.

Like the Romans with the *pelegrini*, the missionaries and traders herded the survivors into a single settlement at Vaihu, so disconnecting their remaining cultural links to their former territory. These people probably only had access to the patchiest histories of what had actually happened since 1720 (if that is when their civilisation was first attacked by Europeans.)

"…after Chile officially annexed the island in 1888,
the few survivors of Rapa Nui's forgotten genocide were

forced into a detention centre in the village of Hangaro, a camp where they were kept confined under the most appalling conditions for nearly 100 years...

'It was surrounded by a barbed-wire enclosure with two gates in it, and no one was allowed to pass through them without the permission of the Chilean military leader. At six in the afternoon these gates were locked...These regulations have remained almost unchanged...In 1964, 1,000 surviving Easter islanders [were] living in the most unbelievable wretchedness and lack of freedom. Maziere, 1969: 35" [254]

Where does that leave us?

So if Easter Island did not collapse due to failure of incest avoidance and Westermark effect, then we have *yet another case* where a Pacific Island population managed to survive for nearly one thousand years. I must admit that I did wonder about the timing of the effect I hypothesized. How long does it take between 40 and 200 people, practising how much incest, to overpopulate an island? This is like asking how long is a piece of string? You have to assign birth-rates, mortality rates, and degree of incest practised. You have to add customs to reduce the occurrence of the Westermarck effect, such as bringing male and female children up separately from birth. There are so many scenarios. It would have been very convenient for the Easter Islanders to have set their reproductive clocks on a suicide course that would occur just before the arrival of the first Europeans.

Even with polygamy, and long life expectancy and low mortality, is it possible to reach populations numbering between 7000 and 10,000 from a base of 100 with only one line of mitochondrial DNA and without severe incest?

It would be surprising if Pacific Islanders had set off in one party for a distant island without making sure that there were enough potential legal partners to seed the new colony, unless they had been fleeing some kind of persecution or natural catastrophe in a hurry. Will we find out, with further investigation, that the initial party was not actually all that closely related?

It is surprisingly difficult to obtain up to date information on these matters.[255]

We are also told that the Easter Island was totally isolated. Was this isolation only from Europeans or must we believe that between 400 AD and 1722 there was no occasion when other Pacific Islanders journeyed to Easter Island? How good is the data for this?

The Pacific Islander 'rule'

The discussion above makes you want to know more about the trade wars and the routes ships took from the 15th century onwards. It makes you wonder about some other tales of collapsed civilisations and want to investigate the dating methods and results and how many may have been affected by 'confirmation bias' where researchers themselves came from societies with land-tenure systems guaranteed to overshoot resources and little awareness of other land-tenure models.

I will investigate some of these matters further on in this volume and more in the second and third books, especially the European Trade Wars and colonization during this period.

In this part I want to give some anthropological detail of the land-tenure and inheritance traditions and the cultural behaviours that assisted the maintenance of a steady-state society.

Easter Island turns out to simply confirm the Pacific Islander steady state rule. Generally, Pacific Islander societies, as we mentioned early in this book, have been the greatest survivors ever known, with some lasting more than 40,000 years.

That people of 'progressive' cultures have so easily identified Pacific Island societies as hopelessly unsustainable lies in the severe dysfunction of most Pacific Islander cultures since the majority were exposed to Progress through colonisation from the 1850s. The spectacular fate of the Easter Island, popularly conceived to have occurred before any colonisation, simply confirms in most *progressive* peoples' minds that the poor Pacific islanders must always have existed in hopelessly dysfunctional societies, doomed to bloom and die like so many tropical blossoms, due to some innate lack of the progress gene.

The faulty system

It turns out that the Pacific Islander Future-eating syndrome[256] belongs to the colonial 'European' model – but not to the Romanised system on the continent. This is the subject of Book Two in this series of *Demography, Territory and Law* series, where we find that the Future-eating syndrome came from the English land-tenure and inheritance system. This system, although it had its origins in France, did not come from France originally.

It came via the Vikings. Moreover although it took root in Britain, it did not persist in France.

But here, in Book One, we will conclude by looking at the qualities that define the steady state system or topology and without which it will not function.

The Steady State system

All of the long-term viable Pacific Islander cultures developed similar land-use and tenure law which kept the Future Eater syndrome at bay because it gave the kind of systemic ecological biofeedback that tells a frog that the water is getting too hot in the pan.[257] Perhaps it could have worked indefinitely if there had been no outside interference.

The key to a stable society seems to lie in the presence of strong incest avoidance structures that prevent the occurrence of children without land-rights and in the absence of commodification of land. The buying and selling of land breaks up kinship organisational structures underlying original societies. Kinship based organisation was preserved in the Pacific Island cultures because the only way of acquiring land was to be born to it. Having said that, you could also be adopted by a family or marry into one and you could be adopted into a clan or tribe as an immigrant. Marriage candidacy was limited by certain taboos and immigration relied on the agreement of the entire clan or tribe that they stood to benefit (usually because the immigrant brought the opportunity for non-incestuous marriage to isolated societies). When you became a member of family, clan or tribe, your position then entitled you to normal inheritance rights as if you had been born to that place and kin, but with those rights came the obligations of true kinship.

In cultures which commodify land, (i.e. buy and sell it to anyone who can pay the price) you can buy land from a member of a community without marrying or carrying over the obligations that went with that land to the rest of the community who would all have had some reciprocal rights or obligations on or around that land. Land defines a community through these rights and obligations. Where these links are broken, the community loses power over the thing that once defined it.

So traditional islanders couldn't sell land,[258] not even to close relatives. That meant that a person could not acquire a large volume of land and then sell it back to other islanders. This limitation took care of land speculation. You also could not sell your land to a foreigner, although you could lease or lend your land to a foreigner if the tribe agreed. This took care of immigration as a source of rapid population growth.

Rapid growth through high immigration occurs where speculators acquire land which has somehow become alienated from its usual local community use. They then reparcel it to carry more people than previously, and sell those parcels off for as much money as possible to people from outside the area. The reason the growth has to come from outside the area is that growth inside the area has limited itself in response to perceived constraints.

In Pacific Islander land tenure and inheritance systems, getting land from another tribe meant that you had to go to war; you couldn't just offer them a lot of money and induce some members of the local community to alienate parts of their land for their individual financial profit whilst depriving the local community of the use of that land which you had once inherited. We will see in Book Two of *Demography, Territory and Law*, on Land Tenure and the Origins of Capitalism, that this disorganization of community with commodification is exactly what happened in Britain. It also happened in France, but not to the same degree, owing to the institution of a different land tenure and inheritance system, which ultimately spread to most parts of continental Europe. (See Book Three.)

How did the Pacific Islander land tenure and inheritance system affect the rate of natural reproduction? If you had too many children your plot of land became unviable and no-one would want to marry into your family because they would be acquiring more mouths to feed with less land or going to live in poverty. The logistics, being subsistence related, were known and the boundaries of the island support system were visible. Land-use rights went by inheritance to legitimate children and traditions evolved to cull the illegitimate, who were defined by being born outside the rules applying to their parents' cast and family.[259]

Transfer of land was restricted to leasing. Land was occasionally lent to foreigners or refugees on condition that it were put to a reasonable use and would revert to the 'real' owner if the lessee didn't use it, and that the lessee would help the 'owner' out with various obligations and tributes. Most islanders remained close to their original homes, but if they left, after a shorter or longer period of absence (depending on the arrangements they may have made to maintain links via relatives or proxies) their lands were traditionally allocated to someone else and they were made to feel unwelcome if they returned.

A feature of Pacific Islander land laws[260] was that the needs of the community took precedence over those of individuals in these subsistence societies. No one person held all the rights to one piece of land; people, family, clan and tribal rights interacted across boundaries. A landholder needed rights to several plots of land so that 15 years fallow could be allowed between many harvests, to prevent soil leaching. In swamps, where some foods like taro could be continuously cultivated, rights were very strictly drawn. Rights to the sea were allocated in a similar complex way, according to distance from the shore, catches, what kind of fish and activities, etc.

Fresh water tended to be owned by whole tribes or clans, with boundaries drawn at the watersheds. Plots of land were often very thin, but drawn from the centre of the island, like pie slices, in order to give the user access to all kinds of land types and features. Land and water rights waxed and waned according to different life phases.

Ultimate entitlement was reserved for people indigenous to the nation, or to a locality within that nation, and strengthened along blood lines. Under stress the rigidity of these rules increased, sometimes to the point of war, which was the other method of land rights redistribution. Stresses included increased population growth or decrease in island productivity. In most, but not all of Polynesia, land rights were mainly patrilineal, but this could vary in some circumstances and islands. In some places there was matrilineal inheritance of specific kinds of land rights.

Pacific Islanders in general seriously lost control of their environment, with the rise of colonialism, and new land-use practices came into being.

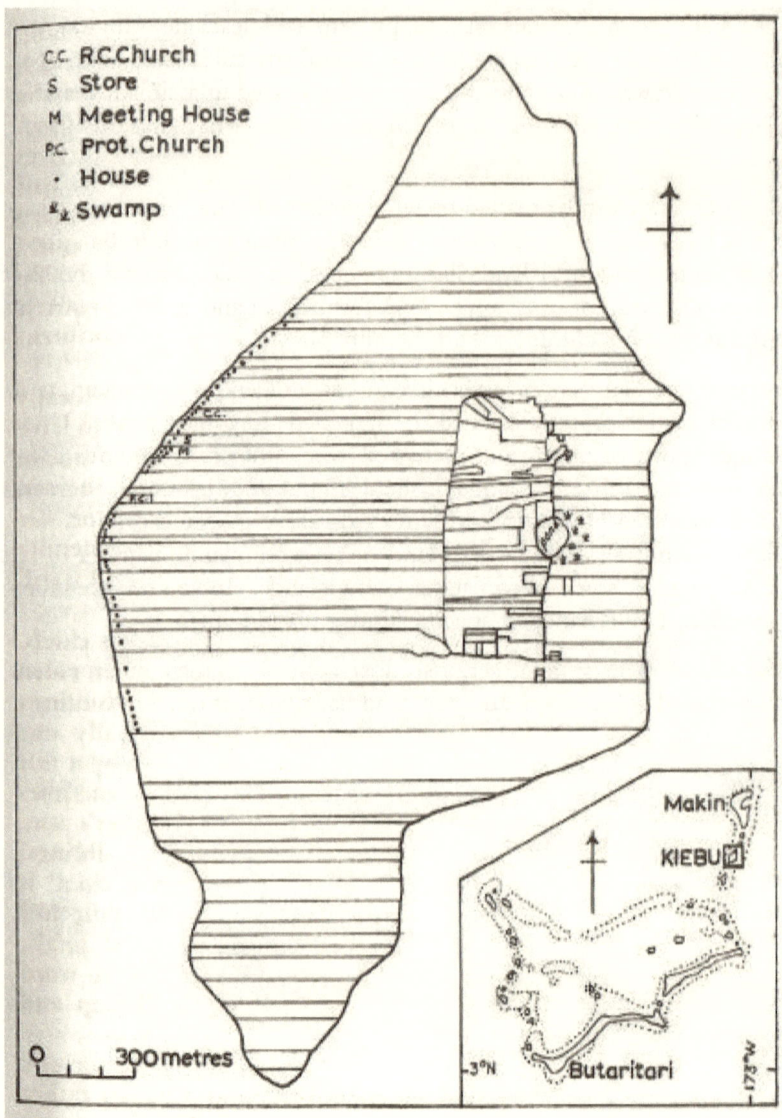

Fig. 2 Kiebu, Gilbert Islands, showing lands of sixty-five ramages
(longitudinal strips are 'forest' land, fragmented plots at centre right
are Cyrtosperma gardens).

Figure 13, Gilbert Islands land-use, 1960s.
Source: Bernard Lambert, "Fig.2. Kiebu," in "The Gilbert Islands," in
Ron Croncombe (Ed.), 1971. *Land Tenure in the Pacific*, OUP. Shows
"pie-slice" way of dividing islands into strips, with the important
crytosperma (giant taro plants) gardens, divided independently of the rest.

The map of the Gilbert Islands land-use in Figure 13 was drawn in the 1960s when there were already two different denomination churches on this tiny island, confusing land-tenure traditions.

Traditional gender role differences among Pacific Islanders were often marked and this contributed to reducing or increasing the rate of natural increase. For instance, common Melanesian Pacific Islander practice was that men and women lived in separate quarters, even if they were married. This tradition was also present if less widespread amongst Polynesian Islanders. Male and female children were often reared separately by the parent of the respective sex - or at least they spent considerable time engaged in separate activities due to gender specificity of tasks. Temporary taboos were also imposed that prohibited certain persons at certain times from having sex, or sleeping and eating with groups. Women would have been less likely to have got pregnant during lactation because they often didn't share the husband's bed at this time. People would not have married and have had children if there was not enough land available to cultivate or fish, since these were subsistence cultures which simply could not afford extra mouths.[261]

We know of a number of such traditions that would have kept population growth down. Some Pacific Islander societies were relatively permissive about casual sex with the Arioi cult a vivid example.[262] Marriage, however, which created offspring with legal claims to land, was subject to strong incest and cast taboos in Pacific Islander cultures and elsewhere.[263] As we know, incest avoidance is almost universal in human societies and it is a population growth regulator. To make this clearer; sex was not the object of taboo in a sexually permissive society; births were.

With regard to the principle of incest-avoidance, there is strong evidence that strictness increases in line with low environmental resources. Among the Mae-Enga – a New Guinea tribe with little wealth, M.J. Meggitt in *The Lineage System of the Mae-Enga of New Guinea*,[264] relates that incest avoidance was effectively practised up to second cousin level, and that this included avoidance of marriage with second cousins in law. Virginia Abernethy in *Population Pressure and Cultural Adjustment* was well aware of the impact of incest avoidance on population growth and reminds us of Birdsell's work on the range of strictness according to climate in Australia:

> "Anthropologists have long marvelled at the complexity of marriage rules in some societies, perhaps the most complicated being the eight-part subsection reported

for many Australian Aborigine tribes. Under this rule only one type of second cousin may be married, and given that the rules are prescribed among tribes living in the sparsely settled central and western desert regions, the impact on marriage and therefore reproduction is potentially great if the rule is followed. ... Thus, it is of interest that the coastal Aborigines, who live in the most productive area of Australia (because of its higher rainfall) have a simpler rule: they choose mates by a moiety (two-part system), according to which only half, rather than seven-eighths of members of the opposite sex are forbidden as marriage partners. In fact, the number of marriage classes rises from two to eight as the rainfall drops to lower values, so that the proportion, and absolute number, of eligible mates is in an inverse relationship with the carrying capacity of the environment..."[265]

Incest and caste taboos can make a lot of difference. Imagine a very small isolated society where a large proportion of the people are already related because they descended from one or two families. In such a society, very few legal marriage combinations would be possible.

A case related by Anne Bolin illustrates such restrictions. On an island with a population of 109 people in 1951, seven of the nine women there who were of marriageable age were forbidden to marry locally. In some islands every young person had to either wait for visitors to their island or go to other islands to find partners.[266]

Abortion and infanticide seem to have been customary in a number of situations. Many of the bigger Polynesian societies had strong feudal-like hierarchies. The first born generally held a higher rank than the subsequent siblings. Although living together between persons of different caste was permitted, marriage was not, and the offspring of such unions, casual or long-term, were strangled in Tahiti. The same solution was reputedly used to maintain the childless status of the members of the Arioi Hawaiian religious cult, ostensibly for reasons of rank and religious function.[267] Since titles, positions, and land could be inherited, mainly via the male lineage, it can be inferred that these taboo-influenced killings derived from traditions related to land and carrying capacity. Who, in a subsistence society, would attempt to raise a child with no land rights?

In the Marquesas, the ratio of 2.5 men to one woman has been attributed to the high rate of female infanticide. This ratio was also reflected by the practice of women having more than one husband. Obviously the chances of population growth were reduced in situations where taboos required that mixed caste offspring be killed, where more females were killed than males, and where two or more men had to share one woman. In Tahiti, to the contrary, polygyny was practised, but only one woman's children could inherit titles and property. Presumably infanticide and abortion would also tend to have been practised by the less favoured woman. In this culture male infanticide is supposed to have been more common, and one can see that this practice would have counted more strongly than female infanticide against population growth in that situation.[268] If this is hard for a person raised in a monogamous society to imagine, consider the fact that a man produces enough sperm to impregnate many women every year. Opportunities are limited by logistics and by taboos. If a man can have several wives then he has the potential to impregnate them all in the same year. In a polyandrous society several men may have only one woman whom they have the right to impregnate, thus strongly curbing their sperm fertility opportunity.

The introduction of cohabitation by married couples to islands where there had been separate male and female tenure and quarters, increasing travel between islands and further afield, the loss of rigid incest and caste taboos, of abortion and infanticide traditions, and the banning of warfare, presumably all had their effect on fertility. Christian religious myths of plenty, cargo-cult phenomena, introduced economic paradigms that ignored the small size of islands, such as western style economic megastructures like mines and banks and monocultures, encouraged the same delusions in the islanders as those most Americans and Australians believed. The risk was of leading island populations to think that their resource horizons were no longer limited to island produce.

These kinds of changes in perception would have meant that the populations grew way beyond the capacity of the islands to provide properly for them and for the land use allocation system to work. A small plot of land sufficient for a family might now be owned by hundreds of persons.[269]

Some of these island cultures were thousands of years old. Their peoples maintained very pleasant and productive environments. Many lived in places we now call island paradises. Some developed to the point of having kings and queens, but few developed beyond a few hundred people.[270] Dispossession by colonists in the 18th and 19th centuries contributed at first to massive population die-offs.[271] For the survivors, changes to living arrangements for married persons, and new attitudes towards contraception, abortion and infanticide, and the buying and selling of land, saw a massive rise in population in the 20th century which rendered land-use allocation laws and inheritance laws almost useless, as hundreds of relatives would fight (and still fight now, as the situation continues to worsen) over land which had once been inherited by many fewer. As well as cultural changes which affected the birth rate, the islanders lost almost all control over immigration and settlement, which their land laws had previously afforded them. New populations from places far away came to stay. In some places they changed the law immediately. In others they had more difficulty. For example, the United Kingdom forbade the sale of land in Fiji and only allowed New Zealand to administer the Cook Islands on condition that land sale was forbidden.[272] The Act of the Treaty of Waitangi 1975 and subsequent amendments in New Zealand established inalienable land for Maoris.[273]

Despite these relatively enlightened laws, commercial land transactions have recently been permitted in various Pacific Islands or are being strongly pushed by organizations like the World Bank and by foreign economic 'specialists', for instance, in the Cook Islands, in Papua New Guinea, and in Easter Island (Rapanui).[274] Pacific Islanders and experienced food and agricultural scientists who understand the importance of the earlier restrictions are desperately fighting these external impositions and influences.[275]

The pressure of foreign values which commodify land, the globalisation via the internet of the real-estate market, and foreign media misrepresentation of local situations brought irresistible pressure to bear on the islands from the outside. Local population pressure, marginalised immigrant groups, and implosion of islander economies under the rigors of the global marketplace continue to erode the Pacific Islander land-tenure system from the inside. It dooms those islanders who still own land to poverty, overpopulation, dispossession and foreign takeover; it gives little or nothing in return, except that it provides a spearhead for outsiders to wedge their presence and transform these societies for short-term individual gain that erodes the common good.

At this point I should anticipate the criticism that some Polynesian and Micronesian societies have been known for unpleasant attitudes and practices, for instance the harshness of the Enga towards women, whose purpose as spouses seems often to have been only as horticultural slaves whose husbands resisted intercourse because of the notion that intimacy with women weakened them. Fijian society, by the account of Thomas Williams in *Fiji and the Fijians*,[276] could be incredibly cruel and unjust with feudal oppression and arbitrariness mixed with half-useful and half-arcane traditions which related back to the viable steady state islander system. In both cases it is hard to know whether this social nastiness existed before European contact. My theory is that the steady-state system was the outcome of the selection of social systems of organisation which privileged long-term survival. Obviously all of us could have the misfortune to have been born into societies ruled by madmen or ordered by religious hierarchies that required mass human sacrifices, or, as the Anglophone societies are currently, ruled by growth-mad media-annointed urban planners.

Summary of Steady State Topology variables:

I will summarise the material above about the Pacific Islander Rule and construct a topology from it. A topology is a model, a mould, a shape (or a system) that may vary but which is always recognisable by the retention of some essential characteristics, without which it will not work. The characteristics of the Pacific Island tenure system can be summarised to the essential ones. These form a topology which can be subjected to many variations but which will not function if it leaves out any of the essential qualities.

The essential qualities or independent variables of the Steady State or Pacific Islander Land-use planning topology seem to be incest avoidance, non-sale of land, and population number control.

These independent variables are achieved through marriage laws outlawing certain combinations and legitimacy laws defining who might survive infancy and who might inherit.

The methods of population numbers control reinforce the impact of incest avoidance and the Westermarck Effect. Traditionally they include gender separation, contraception, abortion and, in the past, legal or unofficially tolerated infanticide. Modern societies do not allow infanticide, officially or unofficially, but, where contraception and abortion are very reliable, infanticide is not an independent variable – i.e. not necessary to the topology. Gender separation is practised to a greater or lesser degree by all societies.

Incest avoidance, non-sale of land, and population number control are reinforced by caste and marriage laws, other inheritance laws, gender specific practices, adoption, immigration and emigration practices. The caste and marriage laws, other inheritance and gender specific practices, adoption, immigration and emigration practices vary enormously and the degree and effect of possible variation within them, is complicated.

What can go wrong.

How this system can be broken and what went wrong is the subject of the second book in this series, which gives the history and characteristics of the British land-use planning and inheritance system and contrasts it with the Roman-law one in continental Europe.

This is the end of Book One, *Demography, Territory and Law.*

The Demography, Territory and Law book series

You have come to the end of Book One, in the four book series, *Demography, Territory and Law*.

The next three books use the theories here to give a dramatically new view of pre history and history, interpreting political events and systems in neither Marxist nor Capitalist ways.

Book Two – on the Origins of Industrial Capitalism in Britain - is like a new history of Britain, its demographics, and its role in the world. Most works analyzing the rise of industrial capitalism assume that capitalism was an inevitable step in an evolutionary path of economic progress, to which all societies would and should naturally aspire. These works tend to ask the question of why other countries 'lagged' behind Britain in developing industrial capitalism. I ask a more basic question – what caused industrial capitalism in Britain? Using the theory of Book One, Book two explores the question of what drove population growth in Britain much faster than elsewhere. It contains quite a bit of ecology on forests, plagues and malaria, but also economics, law, history and Shakespeare, trade wars and alliances between Spain, Portugal, France, Holland, Britain, slaving, colonialism, and the effects on demography, politics and economies of different kinds of fuels. It begins with Neolithic times, then studies the Roman occupation of Britain, analyses subsequent waves of invaders and immigrants, gets to 1066 and goes to the Restoration and the Dutch Capitalist revolution of the 16th century.

Book Three – on Land Tenure and the Origins of Modern Democracy in France - works through the differences between the systems of Britain and France and asks how come Britain had an industrial revolution and France had a democratic revolution? Looking at the preservation and the dislocation of clan links to place and each other in France and Britain, using Urge to Disperse theory, it identifies important societal structures that helped the French to organise a revolution that managed to maintain itself in and out of power within and without various political parties, despite apparent interruption by several monarchies, and despite the execution of many of its leaders from 1789 to 1846. It explores some overarching processes, such as the new banking, bubbles and debts of France and Britain to finance their colonial trade wars, compares the financial and social effects of the dissolution of British monasteries with the state's resumption of monasteries in France over a century and a half later and asks where all that land went. It also looks at the role of ex-Paris Police Chief and Emperor, Napoleon, in the progress of European democracy and the reaction by the Royal houses and Holy Roman Empire in Britain and on the continent. Finally, it observes how Napoleon's Code, as well as revitalizing Roman law, helped to institutionalize a number of revolutionary values throughout continental Europe, reinforcing systemic differences and outcomes between it and most English-speaking countries.

Book Four – is about the world after Napoleon and the rise of industrial capitalism in Britain. It follows the increasing divergence between the Anglophone states and the 'Roman or Napoleonic' ones as Britain's coal-fueled empire became dominant and local social organization was increasingly fragmented in Britain and her colonies. It looks at the spread of industrial capitalism to continental Europe and other countries before, between and after the First and Second World War, noting local and national reactions, including communism and anarchism, in countries including Spain, Russia, Italy, Greece, Yugoslavia, and Arab and South American states and the fate of these movements. It examines the ongoing trade wars - colonial and corporate - as well as the role of coal and oil - through WW1, WW2, the first oil shock, countershocks and the oil wars of the 21st century. Throughout the breakdown of local power, landlessness, and the phenomenon of huge populations accompanying industrial capitalism and communism is commented – as are exceptions to these demographic trends.

The New Demography, Territory and Law Theory relates Political Organisation to Land and Population Organisation

Above I have broadly traced the contents of the series, but not the theory. In this series, *Demography, Territory and Law*, Book One establishes a theory of a system for social organization that preserves populations in steady-state with their environments. The primary aim of the *Demography, Territory and Law* series was to understand why populations had got so huge since the 18th century and did not seem to be slowing down in some places causing increasingly extreme environmental degradation, loss of familiar, loved, and pleasant natural amenities, wildlife and wild landscapes, and threatening survival in the face of inevitable fossil fuel decline. The fact that some states seemed to retain control over their populations provided a hopeful exception to this apparent rule that mankind is doomed to overshoot his resources. I wanted to investigate this difference.

In Book Two, investigations began where the world's first remarkably fast and sustained population growth began – in Britain. The 'steady state population theory' of Book One would expect such population growth to coincide with disruption of clan and tribal organization and relationship to place. Book Two also looks at other possible causes of this massive population growth, including capitalism itself, as well as fossil fuel. Did fossil fuel cause capitalism or did capitalism cause the creation of the technology to use fossil fuel for industrial processes? Did population start to grow in Britain before or after industrial capitalism? What was the role of malaria and plague?

Book Three looks at a comparable country to Britain, just across the Channel – France. France is a great country to compare with Britain because industrial capitalism did not really take off there until after the Second World War. Prior to Britain's coal-fueled industrial revolution, however, France had an economy comparable in many ways. More interestingly, in the 19th century, when its neighbours, Britain and Germany, had rapidly burgeoning populations and industry, France's population grew comparatively slowly. France is, in fact, known for having conservative population growth patterns. Even after the Second World War, when it embarked on a program of population increase, it pulled up short as soon as the first Oil Shock occurred, unlike most Anglophone countries, which seemed to do everything to increase their population growth. One would expect such a history of low population growth in France to coincide with relative stability of clan relationship to place.

Through this investigation of the causes of overpopulation the association of local population displacement and dispersal with injustice, coercion and lack of political power became obvious. The converse, relative stability of local populations was linked to the ability to organize politically at local, departmental and national level in France. This led to a development where Books Two and Three construct and test, by looking at historical events and trends, a theory that democracy is best served by systems that preserve power in local hands. That local power is most resilient when it is held by populations with strong family and clan links to land and locality. The converse seems to be that capitalism is best served by systems that remove power from local hands and which disorganize, disperse and deracinate families and clans from their links to land and locality. Where money, capitalism and industry are turbocharged by fossil-fuel the market takes on a life of its own, beyond human boundaries and the ability to judge and affect impact, in a butterfly effect familiar to chaos theory. One of these effects is overpopulation. Overpopulation is, according to my theory of steady state societies, controllable if populations retain local power and organization. So is capitalism and the market. This has implications for scale of industry, but that can be managed by governments structured on strong local and regional delegations with the ability to control the market – as in France, which only recently passed a law to halt gas fracking – something that the United States and Australia would be incapable of legislating to outlaw.

Book Four looks at more recent trends and predicts future ones, starting from the world 'after Napoleon' and the rise of industrial capitalism in Britain. It follows the increasing divergence between the Anglophone states and the 'Roman or Napoleonic' ones as Britain's coal-fueled empire became dominant and local social organization was increasingly fragmented in Britain and her colonies. It looks at the spread of industrial capitalism to continental Europe and other countries before, between and after the First and Second World War, noting local and national reactions, including communism and anarchism, in countries including Spain, Russia, Italy, Greece, Yugoslavia, and Arab and South American states and the fate of these movements. It examines the ongoing trade wars - colonial and corporate - as well as the role of coal and oil - through WW1, WW2, the first oil shock, countershocks and the oil wars of the 21st century. Throughout the breakdown of local power, landlessness, and the phenomenon of huge populations accompanying industrial capitalism and communism is commented – as are exceptions to these demographic trends.

ENDNOTES

[1] These were initially all one book, but they separated naturally into interesting areas of their own – too much to combine in one volume.

[2] Such as the one in the outer Solomans called Pigeon Island which was the subject of Lucy Irvine, 2001. *Faraway*, Transworld Publishers. Irvine described the traditional system in fascinating detail, along with the perspective of the elderly American woman survivor of a couple who had come to live there many years ago and who had apparently believed that they had purchased their land from the indigenous islanders, whereas the islanders believed that the land had only been leased, leading to decades of mostly friendly, but uneasy, cohabitation between the American family and the other people there.

[3] Fingleton, Jim. (Ed.) 2005 June. Privatising Land in the Pacific, A defence of customary tenures. *Discussion Paper Number 80*, ISSN 1322-5421, published by The Australia Institute, https://www.tai.org.au/file.php?file=DP80.pdf.

[4] This is the subject of continuous international friction, well exemplified by the case of Fiji. Purportedly on the basis of championing democracy, but spearheaded by corporate interests which include the international media, there is a constant pressure on the indigenous Fijians of non-Indian origin, who have a continuous link with the land which may not be bought and sold, but only be acquired through inheritance, to commodfy land and put it on the commercial market. The outcome of such commodification would be for them to lose power and property to peoples who were unfortunately imported by British colonial invaders to work for the British because the Fijians were self-sufficient and did not need or want to do so. Fortunately for the Fijians of non-Indian origin, laws were made to protect their traditional interest in the land in Fiji, of which they retain 83%. This was certainly not the case in Hawaii, for instance, which is largely foreign owned now.

Cooper, George and Daws, Gavan. 2011. *Land and power in Hawaii: the Democratic years*. University of Hawai'i Press, Honolulu, USA, p.3.

"It was true in mid-nineteenth century when whites (haoles) became influential advisors to the Hawaiian monarchs. These haoles were leading members of a white community which, under the prevailing Hawaiian land system, chafed at not being able to buy and sell land and to obtain secure land tenure for businesses and homes. The advisors convinced a Hawaiian king to replace the traditional land system with a western one. By the end of the century through use of the new system haoles controlled, through ownership or lease, the bulk of the Islands' productive land and water rights."

[5] Fingleton, Jim. (Ed.) 2005 June, *op. cit.*

[6] Exceptions such as Julian Simons, now deceased, who saw technology and cleverness as saving humans from resource depletion indefinitely and Ben Wattenburg, who believes that industrialized countries must grow their populations in order to win in a global competition of civilizations. Wattemberg, Ben J., 2004. *Fewer: How the New Demogrpahy of Depopulation Will Shape Our Future.* Ivan R Dee, Chicago. The problem is that

[7] In all societies, this seems to be true.

Joseph Tainter, 1988. *The Collapse of Complex Societies*, Cambridge University Press, p.205. "Hunters and gatherers, as is well, known, collapse into minimal foraging units (families) when resource or social stress make large complex gatherings impossible."

[8] Villanueva, J.E., 2012, October 2004. Family affairs: The two faces of political dynasties, Business World on-line, Manila, http://www.bworldonline.com/content.php?section=12&title=Family-affairs:-The-two-faces-of-political-dynasties-&id=59508. This article on the importance of modern dynasties in United States political institutions and some others, such as Japan and Singapore.

Official Report on British Crown estate, (No author), 2012, June 21. Another record year with £240.2 million profit and an £8.1 billion capital value. http://www.thecrownestate.co.uk/news-media/news/2012/annual-report-and-accounts-2012/

Article http://www.businessinsider.com/worlds-biggest-landowners-2011-3?op=1

McEnery, Thornton, 2011, March 18. The World's 15 Biggest
Landowners. Business Insider, http://www.businessinsider.com/worlds-
biggest-landowners-2011-3?op=1

In inverse order, with acreage estimate provided by the New
Statesman, Mc Energy reports that these are the biggest landowners in the
world. With the interesting exception of the Pope, they are members of
dynasties, although one is not noble:

"14. Sheik Hamad Bin Khalifa of Qatar: Land: Basically... Qatar.
[Since 1995] The King owns the country and all 4,415 miles of it. [...]
13. James, Arthur and John Irving: Land: The approximately 3.6 million
acres of land held in Maine, New Brunswick and Nova Scotia make the
Irvings the largest landowners in those states and provinces. Background:
As heirs to the J.D. Irving Group of Companies fortune, the three
Canadian brothers also inherited thousands of square miles of forest land
that the company uses as paper and pulp materials in one part of their
very diversified business portfolio. [...] 12. King Mswati of Swaziland:
Land: All 6,704 square miles of Swaziland. Background: One of the last
"Absolute Monarchs" in the world, King Mswati quite literally owns all of
his Southern African nation. There are no land titles or deeds as the
national law dictates that the throne is the sole landowner. [...] 11. Emir
of Kuwait: Land: 4.4 million of the roughly 7 million of acres that make
up the nation of Kuwait. Background: The Emir is an inheritor to the
throne of Kuwait and has managed to hold onto the country, and his own
land with periodic help from U.S. intervention. Part of the land that the
Emir does not control is American military property that serves as the
largest American base in the region. [...] 10. King Letsie 111 of Lesotho:
Land: All 11,718 square miles of Lesotho, including the country's
diamond mines. Background: King Letsie is his second stint as Lesotho's
monarch after being replaced by his exiled father and then being deposed
upon his return, then taking the throne again after his old man's death less
than a year later. And while the monarchy in Lesotho is mostly
ceremonial, Letsie retains legal title over the nation's lands. [...] 9. [Since
2008] King Jigme Khesar Namgyel Wangchuck of Bhutan [presides over]:
Land: All 15,000 agriculturally lush and unique square miles of his
Kingdom of Bhutan. [...] 8. King Abdullah II of Jordan: Land: All
35,637 square miles of The Hashemite Kingdom of Jordan, one of the

very few members of both The Arab League and The World Trade Organization. [...] one of The United States' most loyal allies in the region [...] currently facing his own realities of the upheavals throughout the Middle East. [...] 7. King Gyanendra of Nepal: Land: The 57,000 miles that make up Nepal, [...] [including] Mount Everest. Background: Despite the country's move to a Maoist Republic political structure in the 1990's, the royal family retains nominal ownership of all Nepal [and the king is friendly with the socialist government] . [...] 6. Sultan Qaboos of Oman: Land: The mostly desert nation of Oman, which also includes the ports along the Coast of Sur, a major point of entry for trade in the region. [...] Background: Qaboos is the ruler of Oman and the head of its military. [...] 5. King Bhumibhol of Thailand: Land: All, approximately, 128 million acres of Thailand including its tourist-friendly coast line and economically strong cities. Background: [...] King since 1950[...].[...]longest serving monarch in Asia and so popular is he in Thailand that a rumor of decline to his health caused a shock to the nation's financial markets in 2009.[...] 4. King Mohammed VI of Morocco: Land: The 175.6 million acres of Morocco, the historically key point of trading between Europe and Africa[...]. Background: [Inherits] a monarchy that dates back to the year 1036. His ownership of Morocco's lands and his inherited fortune have led to estimates of his entire wealth being estimated at around $2 billion. [...] 3. Pope Benedict: Land: The 110 acres of The Holy See that constitute Vatican City. Also, roughly 177 million more acreage of various lands owned by the Catholic Church throughout the globe, including the hundreds of Vatican embassies that are legally titled to The Holy See as an independent nation. [...] 2. King Abdullah of Saudi Arabia: Land: Absolute royal control over the oil-rich 830,000 square miles of The Kingdom of Saudi Arabia and a total GDP that was estimated at roughly $618 billion for last year. [...] 1. Queen Elizabeth II: Land: 6.6 billion acres of land worldwide including Great Britain, Northern Ireland, Canada, Australia [...] [and] [...] the all-important Falkland Islands. Background: England's third [...] longest serving monarch, Elizabeth II retains royal title over The British Commonwealth[...]. With her 6.6 billion acres, Elizabeth II is [easily] [...] the world's largest landowner, with the closest runner-up (King Abdullah) [...] who controls 547 million, or about 12% of the lands owned by Her Majesty, The Queen. [...]"

[9] Article 809 of the Korean Civil Code, despite changes in 2005. These changes are discussed below in Cho, Mi-Kyung, 1994-1995. Korea: The 1990 Family Law Reform and the improvement of the status of women. 33 U. Louisville J. Fam. L. 431 1994-1995. Also entitled and available in draft form as Mi-Kyung CHO, Mi-Kyung, "Recent Reform of Korean Family Law," Ajou University, Korea, http://www.law2.byu.edu/isfl/saltlakeconference/papers/isflpdfs/CHO.pdf.

"Provisions which came into force on the date of promulgation on March 31, 2005 (No. 5.6.7.8.).

5. Abolition of the Prohibition of Marriage between Parties with Common Surname and Origin of Surname

Before the reform, one of the regrettable effects of the patriarchal family system was embodied in the rule prohibiting marriage between parties whose surname and whose ancestral paternal origin are common.39 However in practice this provision had already lost its validity when the Korean Constitutional Court declared that it was incompatible with the Constitution in 1997.

The following is a summary of this decision made by the Korean Constitutional Court.

'The legislative purpose of the Marriage Prohibition Clause did not fall under the permissible category of restricting individual human rights for "social order" or "public welfare" prescribed in Article 37 (2)41 of the Korean Constitution. Such prohibition also violated the equal protection clause of the Constitution by discriminating against gender, because it applied only to surnames from the same patrilineal blood. Also, the Marriage Prohibition Clause infringed upon the pursuit of happiness, which includes the freedom to choose one's spouse, and was inconsistent with the right to marry guaranteed by Article 36 (1)42 of the Constitution.'

After this decision of the Korean Constitutional Court, the Korean Supreme Court laid down the regulation governing the Family Register, which makes it possible to register a marriage between parties whose surname and ancestral paternal origin are common, if it is proved that they are not collateral blood relatives within the eighth degree of relationship. ..."

"...The following is the reformed version of prohibition of marriage in the Korean Civil Code.

'Korean Civil Code Article 809 [Prohibition of Marriage between Close Relatives]
Marriage may not be allowed between parties whose relationship of blood relative exists within the eighth degree (including the blood relatives for the real-adopted child kept before real-adoption).
Marriage may not be allowed between parties if either of them is or was the spouse of blood relative within the sixth degree of relationship, or if either of them is or was the blood relatives within sixth degree of relationship of the spouse, or if either of them is or was the spouse of blood relatives by affinity within fourth degree of relationship of the spouse.
Marriage may not be allowed between parties whose relationship of blood relative existed within the sixth degree of adoptive parents lineage and within the fourth degree of adoptive parents affinity.'"

[10] Brian, M. 2012, 2 June. This popular website helps Icelandic couples avoid incest. http://thenextweb.com/shareables/2012/02/06/this-popular-icelandic-website-that-helps-avoid-incest/ "When your country has a population of just 300,000 people and it's not a question of whether you are related to someone, just how far back, an Icelandic genealogy website is successfully identifying connections between couples, helping them avoid incest. The website is called Íslendingabók (the Book of Icelanders) and it lists information about the inhabitants of Iceland, dating more than 1,200 years back. It's the result of a collaboration project between a genetics company, and an anti-virus software entrepreneur, and aims to trace all known family connections between Icelandic citizens."

[11] Affinal restrictions is another word for kinship laws forbidding marriage with certain relatives.

[12] This is the subject of the third volume in the series.

[13] A detailed analysis of the French system and its origins is the subject of volume three in this series.

[14] 'Overshoot' refers to the overshooting of resources by a population and was popularized by environmental sociologists, Catton, W. and Dunlap, R., who are justly famous firstly for their article, Environmental Sociology: A New Paradigm, (1978). The American Sociologist 13:41-49 and Catton, later, for his book, *Overshoot : The Ecological Basis of Revolutionary Change* Urbana: University of Illinois Press, 1980.

[15] Burnham, Peter, 2003. *Capitalism: The Concise Oxford Dictionary of Politics*. Oxford University Press, p.64"Karl Polanyi concludes that capitalism did not emerge until the Poor Law Reform Act of 1834. Capital existed in many forms - commercial capital and money-dealing capital - long before industrialization. For this reason the period between the sixteenth and eighteenth centuries is often referred to as the merchant capital phase of capitalism. Industrial capitalism, which Marx dates from the last third of the eighteenth century, finally establishes the domination of the capitalist mode of production. In contrast to liberals, writers in the Marxist tradition understand twentieth-century developments in terms of the movement from the laissez-faire phase of capitalism to the monopoly stage of capitalism. On the basis of Lenin's famous pamphlet, *Imperialism: The Highest Stage of Capitalism*, the monopoly stage is said to exist when: the export of capital alongside the export of commodities becomes of prime importance; banking and industrial capital merge to form finance capital; production and distribution are centralized in huge trusts and cartels; international monopoly combines of capitalists divide up the world into spheres of interest; and national states seek to defend capitalist interests thus perpetuating the likelihood of war (*see also* imperialism).

[16] The Netherlands case will be explored in detail in Book 2 of the *Demography, Territory and Law* series in the chapters related to the Trade Wars from the 14th century to the 19th century.

[17] This phenomenon of dispossessed labourers in Britain is a core subject of the second volume of Demography, Territory and Law.

[18] Mary Nazzal, April 2005. An Environment Destroyed and International Law, p.3. http://www.lawanddevelopment.org/docs/nauru.pdf. "By 1906, the Pacific Islands Company reached an agreement which granted them mining rights in Nauru at which point the company was renamed the Pacific Phosphate Company. It is interesting to note that at this time the Nauruans could already foresee the demise of their island. A traditional song created in 1910 illustrates this point:
    By chance they discovered the heart of my home
    And gave it the name phosphate
    If they were to ship all phosphate from my home
    There will be no place for me to go
    Should this be the plan of the British Commission
    I shall never see my home on the hill.'"

"[...]The Jaluit-Gesselschaft mining rights were transferred to the Pacific Phosphate Company for a cash payment of £2000, £12,500 worth of shares in the Pacific Phosphate Company and a royalty payment for every ton exported. While the Nauruans were not part of any formal agreement, the Germans paid the native landowners a very modest amount per ton of rock removed from their land."

After Germany was defeated, Nauru became part of a new political agenda designed at the 1919 Versailles and Australia occupied and administered the island.

    See also, TED (Trade Environment Database) Case Studies: Case Number: 412, Case Mnemonic: Nauru, Case Name: Phosphate Mining in Nauru, Case Author: Michael E. Pukrop, May, 1997. http://www1.american.edu/ted/NAURU.htm.

[19] Tim Flannery, 1994. *The Future Eaters*, Reed Books, Melbourne.

[20] This was a yahoo list about energy resources that was popular in the late 1990s and which was one of the first run by Jay Hanson in a series of yahoo lists with an international membership. There is an associated Dieoff website at http://www.dieoff.com. Jay Hanson has a facebook page at http://www.facebook.com/JayInHawaii

[21] Sykes, Brian, 2002. The Seven Daughters of Eve: The Science That Reveals Our Genetic Ancestry, Corgi Press, UK.

[22] This slowing was a precautionary response to the prospect of oil depletion, which threatens the survival of the unprecedentedly large post-war populations. See Chapter 7 of Newman, S.M., 2002. *The Growth Lobby and its Absence in Australia and France*. Environmental Sociology Research thesis, Swinburne University, 2002. http://adt.lib.swin.edu.au/public/adt-VSWT20060710.144805/index.html.

Chapters 6 and 7,

[23] Stephen D. Mumford, 1996. *The Life and Death of NSSM 200: How the Destruction of Political Will Doomed a U.S. Population Policy*, Centre for Research on Population and Security, Research Triangle Park, North Carolina, pp.179-352. This study also contains an appendix recommending population growth reduction as a method of combatting fossil fuel depletion.. (Appendix 2.)   NSSM 200 stood for National Security Study Memorandum. This was an interagency study of world population growth, US population growth, and the potential impacts on national security.   In this work evidence was given to support an argument that population policy initiatives failed largely due to interference from a lobby group of Catholic bishops in the United States. See pp.179-352).

Aristide R. Zolberg, 1993. Are the Industrial Countries under siege. In G. Luciani, (Ed), *Migration Policies in Europe and the United States*. Kluwer Academic Publishers, Netherlands, pp.53-82, p. 61 : "In the 1970s, mounting objections by conservative segments of the citizenry to the presence of culturally and often somatically distinct minorities, as well as the oil crisis and ensuing economic crisis, prompted the governments of the industrial countries to undertake a drastic reevaluation of ongoing immigration, but the difficulty of reducing the flows to the desired level, as well as to restoring the status quo, precipitated renewed fear of 'invasion'. In the United States, in the 'stagflation' 1970s, estimates of illegal immigrants escalated to as high as twenty million, on the basis of which it was argued that the nation had 'lost control of its borders'. The major solution proposed was to impose sanctions on employers of unauthorized labour, but this failed of enactment because of resistance by organized business interests, so that in 1979 the Congress established a commission to overhaul the entire immigration system."

[24] Ray Kurzweil, 2005. *The Singularity*, Penguin Group, USA. Review by Robert D. Steele. www.amazon.com/Singularity-Near-Humans-Transcend-Biology/product-reviews/0670033847/ref=pr_all_summary_cm_cr_acr_txt?ie=UTF8&showViewpoints=1.

[25] Energy costs in fuel and materials, plus the social costs of the politics of wealth distribution (piracy, slavery, environmental impoverishment, third world inequality, destruction of whole peoples through colonization, loss of biodiversity, rise in pollution, social engineering etcetera.)

[26] Newman, S.M., 2002. *The Growth Lobby and its Absence in Australia and France*, Environmental Sociology Research thesis, Swinburne University, Victoria.

[27] Hanna, William and Barbera, Joseph. 1985–1987. *The Jetsons*, distributed by Screen Gems (1962–1963) and Worldvision Enterprises. The Jetsons was a cartoon television series about a future high tech society popular in the 1960s and 1970s. There have, of course, been critical responses to this kind of fantasy future. An excellent example was the animated film by Pixar: Andrew Stanton, Jim Morris, *Wal-E*, Walt Disney, 2008. In this film corporations had further reduced the families to simple cohorts of infantilized consumers, without any pretence of independence, living on an artificial asteroid due to Earth having become uninhabitable, except by a surviving cockroach companion to one remaining functioning robot. Eventually the humans – all terribly obese – return to earth to try to produce their own food and life independently and without pollution.

[28] Society is kept in order by inducing people to oscillate among multiple allegiances by encouraging them to seek out and identify with fashionable brands and activities; these are all substitutes for real identity. Human in vast anonymous societies, with no real lives in common with their near relatives, may spend most of their time looking for their lost tribe.

[29] It is also a notion which appears to be shared by many Georgists and possibly some anarchists. Whereas there appear to exist no Georgist or anarchist states, both these movements, but particularly the Georgist movement, have had considerable national and international influence at times and have not disappeared from the horizon.

[30] Herbert Spencer, *Principles of Biology*, 1864.

[31] Edward Hugh, 2006, Rethinking the demographic transition, http://global.economic.perspectives.googlepages.com/rethinking_the_de mographic_.pdf, p.8. (I am grateful to the author for permission to cite from his useful draft.) "Another model of long run population growth would be that originally outlined by Ronald Lee in the 1980s (Lee, 1987, 1988) and subsequently taken up by numerous economists working in the neo-classical growth tradition (Kreme, 1993, Lucas, 2002 Hansen and Prescott, 2002, Galor 2005, Galor and Weil, 2000). In this account Lee combines the Malthusian and the Boserupian (Boserup, 1965) versions of history to generate a model where human agents are constantly forced to generate new technology as the pressures of increasing population push against the limits of resource supply. This account has the virtue of simplicity, and of providing a pretty good fit with the evolution of global population for a period of about 10,000 year s (from the agricultural revolution to the middle of the twentieth century).

It has the strong disadvantage that nothing within the theory either predicts, forsees or is able to account for below replacement fertility. So the theory is fine as a statement of what has happened (a looking backwards capacity), but it does not seem to describe the present reality of the developed world, and does not seem to have what Lakatos would have considered an essential feature for a progressive research programme: the ability to predict interesting new facts (a looking forwards capacity).

[32] Recent examples abound in the articles published under the tag of http://candobetter.net/GrowthLobby. See also Newman, Sheila. 2002. Chapter 6 in The Growth Lobby in Australia and its Absence in France. The Relationship between the Property Development and Housing Industries and Immigration Policy in Australia and France. Swinburne University, Victoria, Australia, http://adt.lib.swin.edu.au/public/adt-VSWT20060710.144805/index.html.

[33] Syvret, Paul, (2009, 23 May) "Damn 'em all," Courier Mail, Brisbane, Australia.

[34] "Oh, and wind farms are out, too, because a lesser known pink-speckled migratory stuttering sea-albatross might inadvertently fly into a whirling turbine. Buggar. Solar is good, they say. Terrific. Try powering a

city with solar panels, which one would think rely on massive amounts of energy in a nasty factory to be produced in the first place, probably using stuff made by a petrochemical plant for their components." Syvret, Paul, (2009, 23 May), *Op Cit.*

[35] Fleming, David. 2007. Why nuclear power cannot be a major energy source. Refereed paper published by Feasta, p.8. http://www.theleaneconomyconnection.net/nuclear/summary.html.

Fleming, David. 2008. The Lean Guide to Nuclear Energy: A Life-Cycle in Trouble, p.1, points 5 and 6. http://www.theleaneconomyconnection.net/nuclear/Nuclear.pdf

[36] Sheila Newman, 2008. Nuclear Fission Power Options. In *The Final Energy Crisis*, Ed.2., Pluto Press, UK, 2008, p.196.

[37] M. Ragreb, *Isotonic Separation and Enrichment, Chapter 10,* Pp 8-12, dated 3/9/2012: https://netfiles.uiuc.edu/mragheb/www/NPRE%20402%20ME%20405%20Nuclear%20Power%20Engineering/Isotopic%20Separation%20and%20Enrichment.pdf

[38] "Heavy water production", Federation of American Scientists, http://www.fas.org/nuke/intro/nuke/heavy.htm

[39] http://www.cns-snc.ca/Bulletin/A_Miller_Heavy_Water.pdf

[40] Sold on line at Cambridge Isotope Laboratories. http://www.isotope.com/cil/products/listproducts.cfm?cat_id=87&market=research (Accessed 23 August 2012.)

[41] Atomic Energy of Canada http://en.wikipedia.org/wiki/Atomic_Energy_of_Canada_Limited, cited by Wikipedia at http://en.wikipedia.org/wiki/Atomic_Energy_of_Canada_Limited.

[42] Paul Syvret, 2009, 23 May, *op. cit.*

[43] Nardinelli, Clark, 2008. Industrial Revolution and the Standard of Living. *The Concise Encyclopaedia of Economics*. Library of Economics and Liberty, 2nd edition. http://www.econlib.org/library/Enc/IndustrialRevolutionandtheStandardofLiving.html

[44] Barbara Freese, 2006. *Coal, a human history*. Arrow Books, Random House London. Chapter 4, p.83.

[45] *Ibid.*, p.82. Forty-two per cent of urban recruits were rejected for service in the Crimean War on account of physical weakness in 1854, and they had already been passed by regional assessors.

[46] Homrighaus, Ruth Ellen. Baby Farming: The Care of Illegitimate Children in England, 1860–1943. Ph.D. diss., 2003. Rev. ed., 2010, at Historytools, <http://www.historytools.org/babyfarming/baby-farming.html>. Chisholm, Hugh, ed. (1911). "Baby-Farming". Encyclopædia Britannica (11th ed.). Cambridge University Press http://en.wikisource.org/wiki/1911_Encyclop%C3%A6dia_Britannica/Baby-Farming. The practice of giving new-borns to milk-nurses who took on more children than they could cope with, or, for a price, adopting them out in response to advertisements in newspapers, was a particular feature of the children of the poor in cities of the 17th, 18th and 19th century, often written about in literature.

[47] Kaplan, H. S., & Gangestad, S. W. 2005. Life history theory and evolutionary psychology. In D. M. Buss (Ed.), *Handbook of evolutionary psychology*, pp. 68-95. New York: Wiley.

Kaplan, H., Gurven, M., Winking, J. 2009, An Evolutionary Theory of Human Lifespan: Embodied Capital and the Human Adaptive Complex, p.52. http://www.anth.ucsb.edu/faculty/gurven/papers/kaplanetal_ch3.pdf. This work was also published as chapter 3 for: *Handbook of Theories of Aging.* (Editors: Bengtson, V., Silverstein, M., Putney, N., Gans, D). Springer. Pp. 39-66.

[48] Demographic statistics can appear to bolster up life-expectancy claims that turn out to be spurious. Frequently cited demographic transition dogma about life expectancy ignores situations where many people died early and the impact of other events – such as pandemics - which knocked out large portions of populations. The Black Plague drastically reduced the average lifespan in Europe, without necessarily reducing the lifespan of those who survived it. Romans are often described as not very long-lived, but infanticide was common practice for much of Roman history, as it probably was in most societies in different eras, some of them not so long ago. The very high mortality of the British working class and the decline in birth rates during the 1890s depression reflect this practice as well as neglect, malnutrition and disease.

In France and in Australia it was not until 1905 that births were registered and prenatal and antenatal care was encouraged. Prior to this there is plenty of evidence of high infanticide rates. In Australia, according to Hicks, Neville, *This Sin and Scandal: Australia's population debate 1891-1911*, ANU, 1978, much of this evidence is to be found in the *NSW Royal Commission into the Decline in birth rates in NSW Report*. See also Newman, S.M., *The Growth Lobby and its Absence*, Chapter 6, http://adt.lib.swin.edu.au/public/adt-VSWT20060710.144805/index.html for my interpretation of the reasons for this Royal Commission, which differs from Hicks's. In France and other countries a big reason for the registering of births and the medicalisation of birth was to prevent women from modifying the size of their families and it occurred in conjunction with suppression of information about and access to contraception and abortion. See Ronsin, Francis, 1998. *La Guerre des ventres*. Seuil, Paris. Obviously, although some Romans and some 19th C British lived long lives, (Barbara Freese, *Coal, a human history*, Random House, 2006, p.81.) the statistical impact of high infant mortality shortens the over-all life expectancy.

[49] Steckel, R. H., 2004. New Light on the "Dark Ages", The Remarkably Tall Stature of Northern European Men during the Medieval Era. *Social Science History* 28(2):211-229; DOI:10.1215/01455532-28-2-211, Abstract.

"Based on a modest sample of skeletons from northern Europe, average heights fell from 173.4 centimeters in the early Middle Ages to a low of roughly 167 centimeters during the seventeenth and eighteenth centuries. Taking the data at face value, this decline of approximately 6.4 centimeters substantially exceeds any prolonged downturns found during industrialization in several countries that have been studied. Significantly, recovery to levels achieved in the early Middle Ages was not attained until the early twentieth century. It is plausible to link the decline in average height to climate deterioration; growing inequality; urbanization and the expansion of trade and commerce, which facilitated the spread of diseases; fluctuations in population size that impinged on nutritional status; the global spread of diseases associated with European expansion and colonization; and conflicts or wars over state building or religion. Because it is reasonable to believe that greater exposure to pathogens accompanied urbanization and industrialization, and there is evidence of climate moderation, increasing efficiency in agriculture, and greater interregional and international trade in foodstuffs, it is plausible to link the reversal of the long-term height decline with dietary improvements."

[50] Steckel, R.H., 2004. "Men from Early Middle Ages were nearly as tall as Modern people," Columbus, Ohio, *Research News*, Ohio State University, http://researchnews.osu.edu/archive/medimen.htm (Last accessed 19 September 2011 on a page published in 2004.) Richard Steckel is a professor of economics.

[51] See Book 2 in this series on the Origin of the Capitalist System.

[52] Doepke, M. 2000. Growth and Fertility in the Long Run, *Mimeo*, University of Chicago, available in reduced form in Doepke, M., 2004, September. "Accounting for Fertility Decline During the Transition to Growth", *Journal of Economic Growth* 9(3), 347-383. In countries where effective labour laws prohibit the employment of children, those children become costly rather than income beneficial. In those countries where working for wages is the main option for survival for many but where child labour is prohibited, then people who rely mainly or uniquely on salary will have fewer children. Doepke hypothesised that fertility falls where policies such as education subsidies and restrictions on child labour affect the opportunity cost of education. He compared South Korea and Brazil, the populations of which had begun to grow rapidly around the same time. They differed in that South Korea had an effective public education system, and strongly enforced child-labor restrictions, whereas Brazil had a weak public education system and poorly enforced anti-child-labor laws. Doepke showed that fertility declined in association with industrialisation in Korea more than in Brazil.

[53] Barbara Freese, 2006, *op.cit.*, p.80.

[54] Munro, John, 2005. Great Britain as the homeland of the industrial revolution, 1750-1815, The Economic History of Modern Europe to 1914, Lecture Topic No.4, Economics 303Y1, 27 September 2005, University of Toronto, http://www.economics.utoronto.ca/ 5/

[55] Barbara Freese, 2006, op.cit., p.33.

[56] The Dutch population and economic growth explosion is examined in detail in Book 2 of this *Demography, Territory and Law* series.

[57] Abernethy, Virginia Deane, 2004, September 1. Not tonight, sweetie; no energy: a neo-Malthusian looks at fossil fuels and fertility, *Worldwatch*.

This article tests the hypothesis that people restrain their fertility when they perceive a decline in economic fortune. Indicators of economic fortune include the affordability of housing and other vital needs. Abernethy's theory is known as the "fertility opportunity theory".

[58] Although the promotion of some economists via the mainstream media's support for global political and financial institutions has slurred the discipline as one for 'yes-men'. See Bichlbaum, Bonanno & Spunkmeyer, *The Yes Men, the true story of the End of the World Trade Organisation*, Penguin Books, London, 2004. Economics was initially the province of profound polymath thinkers like Malthus and Riccardo, and later, sociologist-economists, such as Marx, who questioned the theory of economic systems by examining their actual impact on real populations.

[59] Hugh, Edward, 2006. Rethinking the demographic transition. Draft. (Permission given to cite.) http://global.economic.perspectives.googlepages.com/rethinking_the_de mographic_.pdf, 2006., p.4. "This idea that the demographic transition implies a mortality rate which remains below a declining but ultimately stabilising fertility rate was a core assumption of the original transition theory, and the idea of replacement level homeostasis coupled with slowly rising life expectancy continues to enjoy wide acceptance today, despite the fact that an increasing number of counter-examples to this possibility may be readily identified, and the number of these counter-examples seems set to grow rather than decline over the years to come."

[60] Gilles Pison, 2008, Juin. Forces et faiblesses de la démographie américaine face à l'Europe, Figure 1, *Population & Sociétés*, Numero 446, INED, Paris, p.1.

[61] See Newman, Sheila, 2008. France and Australia after oil. In Newman, Sheila, (Ed), *The Final Energy Crisis*, 2nd Edition, Pluto Press, UK, for a discussion about post fossil-fuel populations in France.

[62] Wattenberg, Ben J., 1987. *The Birth Dearth: What Happens When People in Free Countries Don't Have Enough Babies?* Pharos Books. Wattenberg was a fellow of the conservative American Enterprise Institute and this book is primarily concerned with keeping US numbers up.

See also United Nations, 2001. *Replacement Migration, Is It a Solution to Declining and Ageing Populations?* Population Division, Department of Economic and Social Affairs, United Nations Secretariat ST/ESA/SER.A/206, United Nations Publication, Sales No.

E.01.XIII.19, ISBN 92-1-151362-6.
http://www.un.org/esa/population/publications/migration/migration.ht
m. This report describes its inquiry thus: "United Nations projections
indicate that over the next 50 years, the populations of virtually all
countries of Europe as well as Japan will face population decline and
population ageing. The new challenges of declining and ageing
populations will require comprehensive reassessments of many
established policies and programmes, including those relating to
international migration. Focusing on these two striking and critical
population trends, the report considers replacement migration for eight
low-fertility countries (France, Germany, Italy, Japan, Republic of Korea,
Russian Federation, United Kingdom and United States) and two regions
(Europe and the European Union). Replacement migration refers to the
international migration that a country would need to offset population
decline and population ageing resulting from low fertility and mortality
rates." This speculative work has been ridiculed in Europe by, for
example, Léridon, H. 2000. « Vieillissement démographique et migrations
: quand les Nations Unies veulent remplir le tonneau des Danaïdes »,
Population et Société, INED, no358.

[63] Dyer, Colin, 1978. *Population and Society in 20th century France*, Hodder
and Stoughton, Kent, UK, p.101: "While in 1940 France was the oldest
nation in the world, she had also for many decades possessed by far the
lowest population growth-rate in the world. In strength of numbers, the
French had been growing steadily weaker in relation to their European
neighbours. The Great War had greatly aggravated an already very serious
demographic situation, and by 1940 the French population generally had
become aware of the serious consequences that a second massacre of
French lives could have upon the very continued existence of the nation."

[64] France Prioux, Magali Mazuy, Magali Barbieri, 2010. L'évolution
démographique récente en France : les adultes vivent moins souvent en
couple. INED, Paris.
http://www.ined.fr/fichier/t_telechargement/37336/
telechargement_fichier_fr_publi_pdf1_conjoncture_3_2010.pdf.

"Ce léger repli est dû à la baisse de la fécondité des femmes âgées de
moins de 30 ans ; l'augmentation de la fécondité après 30 ans s'est
poursuivie, mais elle est moins prononcée qu'en 2008 (tableau 2). Depuis
cinq ans, la fécondité après 30 ans a augmenté chaque année en moyenne

de 22 enfants pour 1 000 femmes, tandis qu'elle diminuait de 7 enfants pour 1 000 femmes en dessous de cet âge. L'année 2009, comme 2007, se situe en dessous de cette moyenne après 30 ans (+ 11), et la baisse est au-dessus de la moyenne avant cet âge (− 23) ; c'est surtout entre 20 et 35 ans que la comparaison avec la moyenne des cinq années est relativement moins favorable, car entre 35 et 40 ans, la hausse est presque aussi soutenue qu'en 2008."

[65] Harvey Leibenstein, 1974. "An Interpretation of the Economic Theory of Fertility: Promising Path or Blind Alley?" *Journal of Economic Literature*, 12 (2).

See also, for another source, Bourcier de Carbon, Philippe. 1998, Jan-Feb. Transition ou révolution démographique? Les insuffisances et les implications de la théorie de la transition démographique. 1: les origines [Demographic transition or revolution? The weaknesses and implications of the demographic transition theory. Part 1: Origins]. *Population et Avenir*, No. 636. Paris, France, Abstract: "The development of the concept of the demographic transition by Adolphe Landry before World War II and the development of population studies in the United States in the 1930s. How American scholars, such as Notestein, developed and expanded the concept of the demographic transition originally conceived by Adolphe Landry. The growing concern with the consequences of uncontrolled fertility in the developing world led to the development of the international family planning movement. Bourcier de Carbon concludes that many scholars have pointed out that there is no obvious connection between the rate of population growth and the level of income per head, and have challenged the relevance of family planning programs for efforts to achieve socioeconomic development."

[66] Cordell, Dennis D., Gregory, Joel W., Piché, Victor, 1994. African Historical Demography: The search for a Theoretical Framework, Chapter 1 of Cordell, Dennis D., Gregory, Joel W., (Eds) *African Population and Capitalism: Historical Perspectives*, Univ of Wisconsin Press, p.14.

"[These chapters] also take issue with the simplistic assumptions that African fertility and mortality have always been high and that mortality necessarily declined as a result of European intervention, thus creating the conditions for Africa's much decried 'population explosion.'

This overly simplistic and ahistorical view has provided fertile soil for the propagation of demographic transition theory as the appropriate paradigm - implicit or explicit - for the understanding of African demography in the recent past. The demographic transition is a descriptive model of the historical evolution of some European populations."

[67] Mainstream media publications lobby for growth and are major vectors for the demographic transition as an unquestioned ideology. Below are two examples. In the Economist one, the writer(s) use the term in a highly contrived argument that minimizes the situation of Africa, illustrating the criticism cited above from Cordell, Dennis D., Gregory, Joel W., Piché, Victor, 1994. African Historical Demography: The search for a Theoretical Framework, Chapter 1 of Cordell, Dennis D., Gregory, Joel W., (Eds) *African Population and Capitalism: Historical Perspectives*, Univ of Wisconsin Press, p.14.

The Economist, 2009, August 27. Africa's population: The baby bonanza, Is Africa an exception to the rule that countries reap a "demographic dividend" as they grow richer? http://www.economist.com/node/14302837?story_id=14302837&fsrc=r ss. "Thanks to its demographic transition. Africa will suffer less from these afflictions than it otherwise would. But it cannot remove them altogether, because the continent's population will continue to grow, albeit more slowly. The hunger, poverty and strife this causes could gravely limit the demographic dividend."

Henry Ergas's article cited below trots out the ideology unquestioningly, generalizing it to the whole world 'long lives' a 'near universal phenomenon' etc.

Ergas, Henry, 2011, January 28. Ageing is worth its challenges. *The Australian* newspaper., News Limited.

"In 1800, life expectancy at birth surpassed 35 years in only a few places; in most of the world, including relatively rich countries, it may not have exceeded 25. By the end of the 20th century, long lives were a near universal phenomenon, with life expectancy exceeding 80 years in the most advanced countries and still rising. At the same time, as ever more children survived infancy, global birth rates fell from 46 per thousand population annually to barely 20. Family size, which rose when infant mortality fell, thus reverted to or below its initial levels. But because the drop in birth rates substantially lagged that in death rates, world population increased an unprecedented sevenfold."

[68] Thomas Williams, 1858. *Fiji and the Fijians* (Vol.1), London in new edition with George Stringer Rowe, Ed. 1982. Published by the Fiji Museum, Suva. Reference to canoe makers, p. 71.

[69] Taylor, Paul, 2001. World Energy Map. http://www.nous.org.uk/energy.map.html and http://www.nous.org.uk/BFMAP.html. These are pages about Buckminster Fuller's concepts. Cedomir Kovacev and Ann Morris, 2010. The Buckminster Fuller Institute, http://www.bfi.org/index.html.

[70] It is an important subject of the other three books in the *Demography, Territory and Law* series.

[71] Youngquist, W., *Geodestinies*, 1997. National Book Company, Portland, Oregon, pp.22-23. Fossil fuels contain many more 'slaves than biomass fuels like wood because they burn hotter and for longer, they came in much greater reserves and, using industrial scale organisation, they have been delivered continuously in large quantities very quickly to suburban mankind. Energy slave ratings for various fuels come from estimations of how many men could perform the work you could get out of burning a material to power a machine. Youngquist gives the following estimated equivalents: person power (PP) equals 0.25 horsepower = 186 watts = 635 British thermal units (Btu/Hr); 300 PP in 1/10 lb uranium (in current commercial technology uranium only delivers 2% of its potential energy); 4700 lbs natural gas; 5150 lbs coal and 8,000 lbs petroleum.

Power is a measure of work units in a period of time. Burning fossil fuel gives us chemical energy. Radioactivity is heat energy that comes from radiation at an atomic level that occurs when actual atoms split, and after this process there is an ultra slight loss of total atomic mass in the radioactive material. The sun itself is driven by fusion power, which incorporates a similar loss of mass. That mass is converted to energy. In his $E=mc2$ (Energy = mass x the velocity of light multiplied by itself) Einstein meant that mass and energy are really equivalent. The first law of thermodynamics states that you cannot create energy. The second law of thermodynamics states that when you convert one form of energy into another some energy/mass is lost and the amount of mass lost is equal to the amount of energy lost. (There is no free lunch). The third law of thermodynamics states that the conversion of one form of energy to another always entails increasing instability. (You cannot turn a sausage back into a pig).

[72] Vaclav Smil, 2000. *Energies*, MIT Press, p.111-112. "Intensive cropping provided almost always lower energy returns than extensive cultivation or shifting agriculture. That is why peasants expanded or intensified their cropping only in response to gradually growing populations. This reluctant expansion and intensification was eventually able to support impressively high population densities.... No traditional agriculture could produce enough food to eliminate extensive malnutrition. Peasant reluctance to increase labor inputs in more extensive or more intensive cropping together with the preference for large families minimizing individual workloads has made it difficult to raise average per capita food supply and to avoid recurrent food shortages. Only rising inputs of fossil fuels into cropping – directly to power machinery, and indirectly to produce it and to synthesize chemicals – could support both larger population and higher, and better, average food supply."

[73] Maddison, A. 2000. *Monitoring the World Economy 1820-1992*, OECD, Paris, 2000. Maddison uses the concept of purchasing parity to estimate economic growth. Maddison divides world per capita economic growth (which he breaks down into a variety of regions) into five phases between 1820 and 1992. According to his data and measures, economic growth was the greatest it had ever been between 1950 and 1973. Afficionados of Peak-oil theory and alert observers will have little difficulty noting that this period coincides with massive post war expansion of the petroleum based economy and world population. The decline of the rate of world per capita economic growth coincided with the first and second oil shocks (approx 1973 and 1980). After 1973 per capita economic growth never recovered and unevenness of growth has increased. Per capita petroleum availability also dropped from that time.

Abernethy, V., 2004, September 1. Not tonight Sweetie, No Energy. *Worldwatch*. When per capital petroleum availability dropped, so did fertility on average world wide and in all regions where income expectations declined in comparison to what they had been before.

[74] This statement is based on the ideas that, as the oil-producing developed countries – especially the USA - increased their need for

petroleum and other materials and energy, they have become increasingly reliant on resources beyond their borders. The world oil market which was created after the first oil shock also entailed greater internationalisation of oil and other mineral supplies. Globalisation was probably part of this trend.

[75] Sheila Newman (Ed.), 2008. *The Final Energy Crisis*, 2nd Edition. Pluto UK, pp101-111, p.110

[76] Mackillop & Newman (Eds.), 2005. *The Final Energy Crisis*, Pluto UK, pp. 228-234.

[77] Mackillop & Newman (Eds.), 2005. *The Final Energy Crisis*, Pluto UK, pp 228-232, pp 99-104.

[78] Portugal's population grew rapidly from 1972 onward which was when other countries in Europe drastically reduced their growth. For statistics on Portugal see Instituto Nacional de Estatística 2011. See Wilkinson. R and Pickard, K., 2010. *The Spirit Level*, Penguin, Australia for Portugal's consistently inferior performance compared to other Western European countries. Japan performed better and has much more equitable wealth distribution than the expansive Anglophone economies.

[79] See Maddison, A., "Evidence submitted to Select Committee on Economic Affairs," House of Lords, London, for the Inquiry into "Aspects of the Economics of Climate Change," http://www.ggdc.net/Maddison/, 20 Feb 2005, p.2. "Past Relation between World Economic Growth and Energy Consumption: (3) Tables 4a and 4b compare the growth of world population and GDP with energy use (in terms of both fossil fuels and biomass) from 1820 to 2001. The energy intensity of GDP rose until 1900 (to 0.42 tons of oil equivalent per $1000) and fell in the course of the twentieth century (to 0.27 tons per $1000 in 2001). Per capita energy use at the world level rose about eightfold from 1820 to 2001." My comment on this is that, although the economist Angus Maddison saw that GDP growth was decreasing, he nonetheless anticipated a long period of rapid growth through the 21st century, if not so rapid as in the last century. But he based this on his perception of trends of 'dematerialization' of the economy, which he inferred from trends in quantities of calories of oil-equivalent used, and failed to account for differences in the nature of fuels and their suitability for different uses.

[80] Cleveland, C.J., Kaufmann, R.K. & Stern, D.I., "Aggregation and the Role of Energy in the Economy," *Ecological Economics*, vol.32, issue 2, pp 301-17, also at http://www.bu.edu/cees/people/faculty/cutler/articles/Aggregation_role_of_energy.pdf., cited in Sheila Newman, 2008. Ed., *The Final Energy Crisis*, Pluto Press, UK, pp13-14. "[The authors]test the 'dematerialisation' explanation of conventional economists for claims that economic growth is proceeding well with reduced growth in global oil-production.

Looking at the causal relationship between energy use and GDP from 1947 to 1996, they find, instead, that people and business have not used less fuel, but that they have been more careful about the fuels they choose to do different tasks, choosing cheaper fuels to do low production work and more expensive ones for high returns. Cleveland et al's main indicator of quality is financial price and it seems to be an indicator which performs well in this case. The authors find that the financial cost of fuels is an accurate reflexion of their versatility or adaptability to specific tasks more than is measuring the total calories they embody. (Neither was any relationship found between the quality of 'transformity' (Odum) and the versatility of fuels for human social needs in industrial societies over the period examined. The authors illustrate this with the example of coal: "Users value coal based on its heat content, sulphur content, cost of transportation and other factors that form the complex set of attributes that determine its usefulness relative to other fuels.")

Logistical or engineering factors affect the return on calories. Humans have adapted their social systems and technology to the limitations of fuels and fuel supply. If a fuel must be transported a long way, it makes less sense to use gas, since the cost rises over distance. To keep expensive pipe diameters low, natural gas, for example, needs to be compressed or cooled to LNG for transport through insulated pipes. (This is also a major reason why using hydrogen as an electricity carrier is so awkward.) You would do better by using a solid (like coal) or a liquid (like petroleum) in this circumstance. Whereas petroleum was once more widely used in North America for domestic heating, because it has a higher calorie content than coal, it now tends to be used for more economically productive work. Coal-fired electricity became a heating option, but now the price of coal is rising due to demand and the cost of transporting it to sprawling populations. Cleveland et al found that the more expensive the fuel in dollar terms, the more carefully it was used. The authors also found that, even where fuel was cheap to start with, its cost would increase in line with the financial return it provided to the users, so that their financial measure showed reliability over time."

[81] Newman, Sheila, 2011, April 11, "Fracking democracy,"http://candobetter.net/node/2348. Comments on this article document evolution of laws in France to totally ban gas-fracking as environmentally and socially untenable.

[82] McQuaid, John, 2009, January. Mining the Mountains. Smithsonian Magazine, http://www.smithsonianmag.com/science-nature/Mining-the-Mountain.html#ixzz275C3YDUt

"Explosives and giant machines are destroying Appalachian peaks to obtain coal. In a tiny West Virginia town, residents and the industry fight over a mountain's fate." "Demand for mountaintop coal has been rising quickly, driven by high oil prices, energy-intensive lifestyles in the United States and elsewhere and hungry economies in China and India. The price of central Appalachian coal has nearly tripled since 2006 (the long-term effect on coal pricing of the latest global economic downturn isn't yet known). U.S. coal exports increased by 19 percent in 2007 and were expected to go up by 43 percent in 2008. Virginia-based Massey Energy, responsible for many of Appalachia's mountaintop projects, recently announced plans to sell more coal to China. As demand increases, so does mountaintop removal, the most efficient and most profitable form of coal mining. In West Virginia, mountaintop removal and other kinds of surface mining (including highwall mining, in which machines demolish mountainsides but leave peaks intact) accounted for about 42 percent of all coal extracted in 2007, up from 31 percent a decade earlier. "

[83] For a discussion about costs of building nuclear power supply in Australia from scratch see Newman, Sheila, 2008. "France and Australia," in Sheila Newman (Ed.) The Final Energy Crisis, 2nd edition, pp251-252.

[84] Hughes, Mark, "Global Food Crisis,"Infoplease.com, http://www.infoplease.com/science/environment/global-food-crisis.html#ixzz275Ng024W "Prices for basic foods such as rice, wheat, and corn have risen 83% since 2005. Compared to the first half of 2007, food prices in 2008 have risen even more dramatically: 130% increase for wheat and an 87% increase for soy. Between March 3rd and April 23rd,

2008, the price for a metric ton of rice rose from $460 to $1,000. This almost doubling in price caused riots in Egypt and Haiti. Other nations (Cameroon, Ivory Coast, Mauritania, Ethiopia, Uzbekistan, Yemen, the Philippines, Thailand, Indonesia, and Italy) have experienced violent protests in reaction to the increased cost of food staples. […]Causes of the Food Crisis: Increased demand on the food supply has caused the price of food to rise. The numerous contributors to the rise in cost and the reduction in supply include biofuels, bad weather, the historically high cost of oil and transportation, increased demand for meat and dairy, and population growth."

[85] See for instance, "Historical Crude Oil Prices, 1861- present," Chartbin, http://chartsbin.com/view/oau (Accessed 20 Sept 2012) or "Evolution in the price of oil, 1975 to 2011," Jewish Virtual Library, http://www.jewishvirtuallibrary.org/jsource/US-Israel/oilprice.html (Accessed 20 Sept 2012) These figures show that, compared to a time of cheap accessible oil from the 1930s to the late 1960s, real prices have been high since the 1970s, with some severe peaks. These prices reflect increasing geological scarcity of easily accessible surface oil, plus geopolitical problems as oil producing states attempt to manage their supplies and get the most they can from the market and oil buyers try to induce lower prices.

[86] Wilkinson. R and Pickard, K., 2010. *The Spirit Level*, Penguin, Australia, beg the question of defining countries as rich or poor according to their average wealth. The book points out that, although Australia or the United States might have relatively large GDPs for their total populations, large sections of those societies are politically, materially, socially and physically disempowered and disadvantaged by their relative poverty. The authors imply that these differences drive seriously negative outcomes in almost any area you care to think of.

[87] Wilkinson. R and Pickard, K., 2010. *Op. cit.*, pp.16-17

[88] Veblen, Thorstein, 1899. *The Theory of the Leisure Class*, http://www.gutenberg.org/files/833/833-h/833-h.htm

[89] John C. Caldwell and Bruce K. Caldwell, 2003. "Pretransitional Population Control and Equilibrium," *Population Studies*, Vol. 57, No. 2., pp. 199-215: http://links.jstor.org/sici?sici=0032-4728%28200307%2957%3A2%3C199%3APPCAE%3E2.0.CO%3B2-L.

[90] John C. Caldwell and Bruce K. Caldwell, 2003. "Pretransitional Population Control and Equilibrium," *Population Studies*, Vol. 57, No. 2., pp. 199-215:, Abstract, p.1.

[91] This is also true of a lot of new zoology. David W., 2006. *Extinction and biogeography of Tropical Pacific Birds*, Univ. Chicago Press, Chicago, London, 2006, p.510: "For two decades now, abundant evidence has been gathered for prehistoric anthropogenic extinction of insular vertebrates. Unwilling to consider timescales longer than decades, most biogeographers still analyze modern distributions as if they were natural. It is disheartening to see pertinent prehistoric information ignored time after time..., whether in the Pacific or any other sest of islands. It is just as disappointing when authors do cite a paper or two on prehistoric extinction but then ignore them in their analyses. Physics envy, simplistic hypotheses, and passion for universal explanation prevail over careful field and laboratory observations, solid data, humility about your own abilities and appreciation for information from other disciplines. I suspect that detachment from fieldwork and thus from the organisms themselves, as well as an obsession with computers and statistics, is partly to blame. With enough manipulation, some sort of 'pattern' can emerge from nearly any data set, especially if you did not collect it yourself and are unaware of its limitations."

[92] Peter Pirie, 2000, June. Untangling the Myths and Realities of Fertility and Mortality in the Pacific Islands, *Asia-Pacific Population Journal*.

[93] Peter Pirie, *Op Cit.*, p.15.

[94] Ibid..

[95] Birdsell, Joseph B. 1971. Australia: Ecology, spacing mechanisms and adaptive behaviour in aboriginal land tenure, in Ron Crocombe, (Ed.), *Land Tenure in the Pacific*, OUP/MUP 1971, pp.334-361

[96] Peter Pirie, *Op Cit.*, p. 13.

[97] Ibid..

[98] Ibid.

[99] These population problems are not marked in French territories where the locals benefit from the same inheritance rules and land-use planning as the mainland French population. Japanese islands in the Pacific (Japan has an archipelago of 6,852 islands) also do not have the

same overpopulation problems as those colonized by the British, the Americans or the Spanish.

[100] Peter Pirie, *Op Cit.*, p.13.

[101] Ibid.

[102] Peter Pirie, *Op Cit.*, p.14.

[103] *Ibid.*. "There seems to have been a hunting and gathering stratum throughout the sequence of settlement in the Pacific. This was certainly true of the Maori who, late in the sequence, settled one of the most marginal areas of Polynesia. Although derived from an agricultural society in tropical Polynesia, the Maori who settled in non-tropical New Zealand were content in the early stages of their settlement, when population density was very low and feral resources plentiful, to depend upon a hunting and gathering mode of existence that was semi-nomadic. Only after about 1300 AD, when the large, flightless birds known collectively as "moa" had been rendered virtually extinct, the seal resources seriously overexploited and the Andean kumara (sweet potato) had been introduced, did the culture shift to the primarily agricultural mode of the classic Maori. In time, the economic advantage shifted to the mesothermal climates of the North Island, but the archaic culture persisted in the South Island until after contact (Bellwood,1979:387).

This matter is of great demographic interest, because of the low fertility always associated with mobile hunting and gathering bands. It has been noted that, among these groups, replacement is kept deliberately low and various practices are employed, with varying degrees of emphasis, to lower average family size. These include post-partum and other taboos, prolonged lactation, customs such as bride-price which delayed marriage, abstinence in marriage, the deprecation of sexual interest, attempted contraception, abortion and infanticide. All these practices have been observed in Papua New Guinea among peoples more recently contacted, most of whom are currently sedentary agriculturists (Buhner, 1971; McDowell, 1988)."

[104] Ibid.

[105] Ibid.

[106] Ibid.

[107] Peter Pirie, 2000, June. p.10.

[108] *Ibid.*

[109] Kaplan, H., Gurven, M., Winking, J. 2009, "An Evolutionary Theory of Human Lifespan: Embodied Capital and the Human Adaptive Complex,". For: Handbook of Theories of Aging. (Editors: Bengtson, V., Silverstein, M., Putney, N., Gans, D). Springer. Pp. 39-66. Also available at http://www.anth.ucsb.edu/faculty/gurven/papers/kaplanetal_ch3.pdf.

[110] Kaplan *Op.cit.*, p.40.

[111] Ibid.

[112] Ryan, Frank, 2009, *Virolution*, Collins, London.

[113] Kaplan et al 2009, *op.cit.*, p.42.

[114] Kaplan, *Op Cit.*, p.49.

[115] Dilworth, C., 2010. *Too Smart for our Own Good: the ecological predicament of humankind*, Cambridge University Press, New York; Hopfenberg, R., Hopfenberg, E., and Salmony, S., 2011,"The Expansion of the Classic Demographic Transition Model" (Powerpoint file), Global Population Speakout, http://www.panearth.org/.

[116] Hopfenberg, R., Hopfenberg, E., and Salmony, S., 2011,"The Expansion of the Classic Demographic Transition Model" (Powerpoint file), Global Population Speakout, http://www.panearth.org/.

[117] Tim Flannery, *The Future Eaters*, Reed Books, NSW, 1994.

[118] Dilworth, C., 2010, p.89.

[119] Dilworth, C., "Overpopulation and the Vicious Circle Principle----by Craig Dilworth," January 29th, 2012, http://candobetter.net/node/2755 (Accessed 27 August 2012) "Next are the **sexual** instincts, found in all sexually reproducing animals, first among them being to impregnate or get impregnated."

[120] Dilworth, C., 2010, pp60-70, 288. He surmises a complementarity between female infanticide and male fighting deaths whereby young males tend to outnumber young women until the male numbers dwindle through violence to female levels.

[121] Dilworth, C., 2010, pp227-228. See also pp.108, 277,

[122] Dilworth, C., 2010, pp 58-60 he mentions various traditions affecting fertility opportunities and p. 288 he mentions the practice of delaying marriage ages and joining monasteries and nunneries.

[123] Dilworth, C., 2010, pp 44 and 61. He does make a references to the possible impact of territoriality in limiting breeding pairs by limiting

space, assuming that this to be primarily a function of competition and fitness.

[124] Dilworth, C., "Overpopulation and the Vicious Circle Principle---- by Craig Dilworth," January 29th, 2012, http://candobetter.net/node/2755 (Accessed 27 August 2012) What limits the growth of a population, preventing disequilibrium due to overpopulation, are various sorts of **check**. The most fundamental of these are **external checks**, though physically internal checks such as bodily aging are also fundamental. External checks stem from outside the population, and are typical of r-selected species. They can take the form of a limit of some kind – e.g. of food or breeding sites; or they can take the form of disease, or predators.

[125] Malthus, Thomas, 1826. Of the Checks to Population in Switzerland in Of the Checks to Population in the different states of Modern Europe in *An Essay on the Principle of Population, Book II* 6th edition, Library of Economics and Liberty, http://www.econlib.org/library/Malthus/malPlong7.html#II.I.9

Here Malthus discusses European steady-state societies. "In the parish of Leyzin, noticed by M. Muret, all these circumstances appear to have been combined in an unusual degree. Its situation in the Alps, but yet not too high, gave it probably the most pure and salubrious air; and the employment of the people, being all pastoral, were consequently of the most healthy nature. From the calculations of M. Muret, the accuracy of which there is no reason to doubt, the probability of life in this parish appeared to be so extraordinarily high as 61 years. And the average number of the births being for a period of 30 years almost accurately equal to the number of deaths, clearly proved that the habits of the people had not led them to emigrate, and that the resources of the parish for the support of population had remained nearly stationary. We are warranted therefore in concluding, that the pastures were limited, and could not easily be increased either in quantity or quality. The number of cattle, which could be kept upon them, would of course be limited; and in the same manner the number of persons required for the care of these cattle.

II.V.20

Under such circumstances, how would it be possible for the young men who had reached the age of puberty, to leave their fathers' houses and marry, till an employment of herdsman, dairyman, or something of the kind, became vacant by death? And as, from the extreme healthiness of the people, this must happen very slowly, it is evident that the majority of them must wait during a great part of their youth in their bachelor state, or run the most obvious risk of starving themselves and their families. The case is still stronger than in Norway, and receives a particular precision from the circumstance of the births and deaths being so nearly equal."

[126] This argument should not negate other valid theories, which could be complementary, such as the discovery of fire and spears and the associated ability to vastly increase hunting success, which may have induced food-scarcity before, during or after land-shortages which may have occurred through climate change.

[127] Hobbes, Thomas, Hobbes, 1909-14 Leviathan XIII "Chapter XIII.: Of the Natural Condition of Mankind As Concerning Their Felicity, and Misery." *Leviathan.*
http://oregonstate.edu/instruct/phl302/texts/hobbes/leviathan-c.html#CHAPTERXIII

"In such condition, there is no place for industry; because the fruit thereof is uncertain: and consequently no culture of the earth; no navigation, nor use of the commodities that may be imported by sea; no commodious building; no instruments of moving, and removing, such things as require much force; no knowledge of the face of the earth; no account of time; no arts; no letters; no society; and which is worst of all, continual fear, and danger of violent death; and the life of man, solitary, poor, nasty, brutish, and short."

[128] Engels, 1971:119, in Martha E. Gimenez, 1971. *The Population Issue: Marx vs Malthus,.*
http://www.colorado.edu/Sociology/gimenez/work/popissue.html.

"There is, of course, the abstract possibility that the number of people will become so great that limits will have to be set to their increase. But if at some stage communist society finds itself obligated to regulate the production of human beings, just as it has already come to regulate the production of things, it will be precisely this society and this society alone which will carry this out without difficulty."

[129] Kates, Carol A., 2004. *Reproductive liberty and overpopulation.* Department of Philosophy, Ithaca College
http://www.ithaca.edu/hs/philrel/replib.pdf
"Locke did not consider the possibility that overpopulation might create a conflict between the natural purpose of the family (hence, liberty to procreate) and the preservation of human life."
John Dunn, 1982. *The Political Thought of John Locke: An Historical Account of the Argument*, Cambridge University Press, USA, Australia, paperback edition. (First edition 1969.). pp.118-119 and p. 160.
Lock believed that, "External conflict, arising from the land hunger caused by economic development and population growth, combines with internal conflict to make government essential for the maintenance of internal order and the direction to best advantage of the external protective power of the society." (p.118) ... "There does remain a sort of primitivism in Locke's politics but it is the primitivism of a man who knows that there is no return – and it is a primitivism which is altogether more effectively assuaged by the first of these historical recompenses than it is by the glories of English constitutionalism."
[130] In agriculture, for instance, in paddocks, or feedlot farms, or in pet animals, in houses and yards. Domesticated social animals are prevented from normal sexual and familial associations.
[131] Kaplan, H. S., & Gangestad, S. W. 2005. Life history theory and evolutionary psychology. In D. M. Buss (Ed.), *Handbook of evolutionary psychology*, pp. 68-95. New York: Wiley.
[132] Marx, K. and Engels, F. 1998. *The Communist Manifesto.* (First edition 1883). Penguin, New York.
[133] Bakunin, Michael, Bakunin in The Program of the Brotherhood (1865) as published in *God and the State, No Gods, No Masters* Vol 1, p133 – 137, cited in *Red & Black Revolution.* No. 6. 2002, Winter at
http://flag.blackened.net/revolt/rbr/rbr6/bakunin.html "The advent of liberty is incompatible with the existence of States.......the free human society may arise at last, no longer organised .... from the top down.... but rather starting from the free individual and the free association and autonomous commune, from the bottom up...."
[134] Locke, John. 1821. *Two Treatises on Government.* R. Butler, London.

[135] Émile Durkheim, 1967. *La division sociale du travail.* 8th Edition. (First edition 1893).Presses Universitaires de France, Paris.

[136] For instance, Marx & Engels, 1998.*The Communist Manifesto*, (First edition 1883) Penguin, New York.

[137] This indirect relationship with the land in complex societies is not unique to humans. It exists in complex insect societies, for instance. In ecological communities there are very complex relationships between species, such as in coral reefs. In coral reefs there is also arguably a built environment that makes use of natural systems by pasting and building from organic secretions.

[138] In an industrial society, which relies on many workers and consumers, the workers have the function to produce and consume. The only way that they could exit this system would be to live off their own land, but landlessness is a major criteria for the creation of the worker/consumer system. Land parcels are kept too small, too high priced and locked into a costly service grid. This structure prevents independence. It might be argued that workers could invest in the share market, but they still have to earn and save from a wage to do this

[139] Doepke, M. 2000. Growth and Fertility in the Long Run, Mimeo, University of Chicago, available in reduced form in Doepke, M. "Accounting for Fertility Decline During the Transition to Growth", *Journal of Economic Growth*, 9(3), 347-383, September 2004.

[140] Newton Freire-Maia, "Frequencies of Consanguineous Marriages in Brazilian Populations," *American Journal of Human Genetics*, 4:194-203, 1952. This article records high frequency of cousin marriage and uncle/aunt marriage to niece/nephew in isolated populations in Brazil, a practice which decreased in the 20th century if those populations became less isolated, but which persists among certain classes and 'races', due to culture and remains higher than in Europe. The high rate of introduced populations (slave-immigrant) and the high risk of slaves and freemen not knowing their parentage made a proportion of marriages unverifyable. Also, Dain Edward Borges, *The family in Bahia, Brazil, 1870-1945*, Stanford Univerity Press, California, 1992, p.125

[141] Article 809 of the Korean Civil Code, despite changes in 2005. These changes are discussed below in Mi-Kyung Cho, "The Family Law Reform and the Improvement of the Status of Women" 33 University of

Louisville Journal of Family Law (Annual Survey of Family Law) 437, also entitled and available in draft form as Mi-Kyung CHO, Mi-Kyung, "Recent Reform of Korean Family Law," Ajou University, Korea, http://www.law2.byu.edu/isfl/saltlakeconference/papers/isflpdfs/CHO. pdf.

Provisions which came into force on the date of promulgation on March 31, 2005 (No. 5.6.7.8.).

"5. Abolition of the Prohibition of Marriage between Parties with Common Surname and Origin of Surname

Before the reform, one of the regrettable effects of the patriarchal family system was embodied in the rule prohibiting marriage between parties whose surname and whose ancestral paternal origin are common.39 However in practice this provision had already lost its validity when the Korean Constitutional Court declared that it was incompatible with the Constitution in 1997.

The following is a summary of this decision made by the Korean Constitutional Court.

'The legislative purpose of the Marriage Prohibition Clause did not fall under the permissible category of restricting individual human rights for "social order" or "public welfare" prescribed in Article 37 (2)41 of the Korean Constitution. Such prohibition also violated the equal protection clause of the Constitution by discriminating against gender, because it applied only to surnames from the same patrilineal blood. Also, the Marriage Prohibition Clause infringed upon the pursuit of happiness, which includes the freedom to choose one's spouse, and was inconsistent with the right to marry guaranteed by Article 36 (1)42 of the Constitution.'

After this decision of the Korean Constitutional Court, the Korean Supreme Court laid down the regulation governing the Family Register, which makes it possible to register a marriage between parties whose surname and ancestral paternal origin are common, if it is proved that they are not collateral blood relatives within the eighth degree of relationship. ..."

"...The following is the reformed version of prohibition of marriage in the Korean Civil Code.

'Korean Civil Code Article 809 [Prohibition of Marriage between Close Relatives]
    Marriage may not be allowed between parties whose relationship of blood relative exists within the eighth degree (including the blood relatives for the real-adopted child kept before real-adoption).
    Marriage may not be allowed between parties if either of them is or was the spouse of blood relative within the sixth degree of relationship, or if either of them is or was the blood relatives within sixth degree of relationship of the spouse, or if either of them is or was the spouse of blood relatives by affinity within fourth degree of relationship of the spouse.
    Marriage may not be allowed between parties whose relationship of blood relative existed within the sixth degree of adoptive parents lineage and within the fourth degree of adoptive parents affinity.'
    [142] See Book 2 of this series for discussion about history of women's right to manage land in France.
    [143] Unger, J., 1943, Dec. The Inheritance Act and the Family. *The Modern Law Review*. Vol.6. Issue 4. First published online 18 Jan 2011. p.22: "Finally, the origin of the modem system of intestate succession established under the Administration of Estates Act, 1925, may be cited in support of the contention that the power of testation was not regarded in English law as hostile to the family. (...)The change effected in 1925 consisted in the adoption of a uniform system of intestate succession which revealed the decline in the economic importance of realty and in the social importance of marriage settlements of land, as the feudal rules of descent of nalty had been adapted to the needs of mamage settlements by the ingenuity of conveyancers. The change also consisted in improvement of the rules of succession, consisting, as regards realty, in the abolition of primogeniture, and as regards personalty, in modifications of the system enacted in the Statutes of Distribution which were required by changes in the structiire of the family."

Falsey, Marie, 1985. Comments: Spousal Disinheritance: The New York Solution – A Critique of forced share legislation. *Western New England Law Review*, Vol. 7 (1984-1985) Issue 4 (1985), Note 5. http://assets.wne.edu/160/48_comm_Spousal_.pdf, "Parliament abolished the law of primogeniture centuries later with the administration of Estates Act, 1925, IS & 16 Geor. 5, ch. 23, § 45."

[144] For an update see, for instance, "An Introductory Guide to French succession law and French inheritance tax," Chez Riviera Real Estate, http://www.chezriviera.com/inheritancelaw.html. (Accessed 23 August 2012.)

[145] Abernethy, Virginia Deane, 2004, September 1. *Op. cit.*

[146] Abernethy, Virginia Deane, 1999. *Population Politics.* Transaction Publishers. New Brunswick USA and London, UK. pp46-47.

[147] Goldstein, Melvyn C. When Brothers Share a Wife. *Natural History.* 96(3):109-112, 1987. http://anthropologyman.com/files/15_When_Brothers_Share_a_Wife.pdf

[148] *Ibid.* According to this article, a number of these women still have children from discreet affairs but they tend to have fewer children than married women.

[149] Annie George, 2008. "Widowhood in India," Demography, University of Kerala, Powerpoint production, http://www.slideshare.net/guestb82083/widowhood-in-india. (Accessed 27 August 2012).

[150] Danka Ledgerwood, 2005. Happiness in Micronesia. Palau. http://www.dankainmicronesia.com/history.html

Flood,Bo, Strong, Beret E., Flood, William, 1999. Pacific Island legends: tales from Micronesia, Melanesia, Polynesia, and Australia. the Bess Press, USA., p.94.

"In certain rural areas, men and women live in separate houses. They live separately out of custom and respect for the differences between them." "The role of most of these houses, reserved for men, is to assure the relation between the ancestral spirits and the ... separated from that of the men. It is usual to find separate paths and buildings reserved exclusively for women." Oliver, Paul, 1997. Encyclopedia of vernacular architecture of the world: Cultures. Volume 2, Cambridge University Press, p.1151.

Daley, Caroline and Nolan, Melanie, (Eds.) 1994. *Suffrage and Beyond: International Feminist Perspectives.* Aukland University Press, p.111, "In Melanesia, as in Polynesia, Christian missionaries were concerned about the absence of connubial domestic life in many Melanesian societies. In such societies men and women lived apart a great deal of the time, with men spending much of their time in the communal men's house, and women in their separate households. Through their evangelical activities the churches tried to substitute new models of family life in place of traditional ones."

[151] Abernethy, Virginia Deane, 1999. *Op Cit.*, pp49-51.

[152] Many sources, including, Katherine Blocksdorf, "How old is the oldest horse?", http://horses.about.com/od/understandinghorses/a/How-Old-Is-The-Oldest-Horse.htm (Accessed 27 August 2012.)

[153] Newman, Sheila, *The Growth Lobby & its absence.* Chapter 7, http://adt.lib.swin.edu.au/public/adt-SWT20060710.144805/index.html

[154] Malthus, Thomas, 1826. Of the Checks to Population in Switzerland in Of the Checks to Population in the different states of Modern Europe in *An Essay on the Principle of Population, Book II* 6th edition, Library of Economics and Liberty, http://www.econlib.org/library/Malthus/malPlong7.html#II.I.9. *Op cit.*

Malthus, the classic whipping-boy for Marxists and a poster-boy for many biological scientists, is usually given credit for his earlier work (1798) where he saw no way out of the Hobbsian destiny apart from abstinence on the part of the numerous poor. However Malthus did not stop his studies there. He did a world tour, inquiring into births, deaths, marriages and making statistical notes, published in 1826. In the extract here, he discusses European steady-state societies. Thomas Malthus, 1826.

[155] Drummond, Lorna. 1973. Social Control by Subsistence Patterns in Sub-Saharan Africa. Honors thesis for degree in Anthropology, Wichita State University http://soar.wichita.edu/dspace/bitstream/handle/10057/1672/LAJ_v.6_p.57-109.pdf?sequence=1;

On the kinship and marriage rules of the !Kung: Nancy Howell, 2009. Demography of the Dobe !Kung. Transaction publishers, New Jersey. First published 1979. p. 278, p.306;

Lephoto, Catherine, Yarnell, Jacob, 1996, December 16th.Marriage and Reproduction in !Kung Societies. In About the !Kung San of Western Botswana. A Lawrence University Anthropology Page. http://korea.gnu.org/people/chsong/cb/homesteading/kung_marriages. html.

[156] Homrighaus, Ruth Ellen. Baby Farming: The Care of Illegitimate Children in England, 1860–1943. Ph.D. diss., 2003. Rev. ed., 2010, at Historytools, <http://www.historytools.org/babyfarming/baby-farming.html>. Chisholm, Hugh, ed. (1911). "Baby-Farming". Encyclopædia Britannica (11th ed.). Cambridge University Press http://en.wikisource.org/wiki/1911_Encyclop%C3%A6dia_Britannica/ Baby-Farming. The practice of giving new-borns to milk-nurses who took on more children than they could cope with, or, for a price, adopting them out in response to advertisements in newspapers, was a particular feature of the children of the poor in cities of the 17th, 18th and 19th century, often written about in literature.

[157] Dilworth, C., 2011, p.247.

[158] In the case of China, here is a potted history that describes a thousand or more years of relative stability, followed by extremely rapid population growth. Asia for Educators, 2009. "Issues and Trends in China's Demographic History," Columbia University, http://afe.easia.columbia.edu/special/china_1950_population.htm: "As early as 2 C.E. during the Han dynasty, China had a population of some 60 million — approximately one-fourth of the world's population at that time. Historical fluctuations of growth and decline kept dynastic China's population between 37 and 60 million over a period of at least the next 1000 years before beginning to increase rapidly. In the early years of the Ming dynasty in the late fourteenth century, China's population began dramatic changes that continue to the present. Rapid increases occurred especially between 1749 and 1811 during the Qing dynasty when the country's population doubled from 177,495,000 to 358,610,000. By 1851, the population reached perhaps 431,896,000 before the effects of the disastrous Taiping Rebellion brought about a slowing of past growth patterns (Some 30,000,000 deaths occurred between 1851-1864 during the

upheavals associated with the attempt to establish the Taiping Heavenly Kingdom. In some areas of central China, the effects of this were not reversed until the mid-twentieth century)."

[159] New South Wales Government, Report of the 1804 Royal Commission into the "Decline in the New South Wales Birth Rate," described in Hicks, Neville, 1978. *This sin and scandal: Australia's population debate 1891-1911*, ANU. Only a few of the actual reports were ever published and they were made very difficult for the public to access because they contained contraceptive and abortion information that was outlawed shortly after the Inquiry.

[160] Drummond, Lorna. 1973. *Op cit.* p. 278, p.306.

Lephoto, Catherine, Yarnell, Jacob, 1996, December 16th.Marriage and Reproduction in !Kung Societies. In About the !Kung San of Western Botswana. A Lawrence University Anthropology Page. http://korea.gnu.org/people/chsong/cb/homesteading/kung_marriages.html.

[161] "Algorithm": See beginning of next chapter for a definition.

[162] The ideas in this chapter were first published in Newman, Sheila, 2011. The Urge to Disperse: A new evolutionary theory of the Westermarck Effect and Incest Avoidance in population numbers and land-use planning, Candobetter Press, Australia. http://www.lulu.com/shop/sheila-newman/the-urge-to-disperse/paperback/product-16253208.html.

[163] Source: E.O. Wilson, "Nature Matters", *American Journal of Preventative Medicine*, (2001), Apr; 20(3):241-2. Some other examples of epigenetic processes are heritable changes in gene function that occur without a change in sequence of DNA. Two examples are 'X inactivation' where one X chromosome becomes inactive in females, and 'gene silencing'.

[164] Daadoun, Roger, 1999. "Les régulations sociales de la sexualité", *Encyclopedie Universalis*, Electronic Edition.

[165] (Discussed further on.) This research is summarised in: Dallwig, Rebecca, 2009, November 1. The Common Marmoset, Callithrix jacchus, Current Research, p 8, III. Biomedical Research [cont.], http://pin.primate.wisc.edu/callicam/research8.html and appears in many articles, including:

Abbott,D. H., Saltzman, W., Schultz-Darken, N. J., & Tannenbaum, P. L. (1998);

Abbott, D. H.,Saltzman, W., Schultz-Darken, N. J., & Smith, T. E. (1997);

Baker, J. V., Abbott, D. H., & Saltzman, W. (1999).

[166] The social learning theory of incest taboos is also challenged by the theory that incest is more likely to breed-in healthy chromosomes and to breed-out defects than to compound ever more complex defects. This would be because, where genetically healthy relatives interbreed, they will not have unhealthy offspring. Where relatives carrying identical but recessive unhealthy chromosomes interbreed, their offspring are likely to be unhealthy with reduced survival and reproductive chances. The reduced reproductive chances could result in the breeding out of the problematic genes from the gene pool. It seems likely to me that the balance between endogamy and exogamy in clans within tribes probably also balances these factors, but I have not looked closely at this.

[167] Lévi-Strauss, Claude, 1967 *Les structures élémentaires de la parenté* (*Elementary Structures of Kinship*). Mouton et Maison des sciences de l'Homme, Paris, La Haye. (First edition 1947.)

[168] Ibid.

[169] For instance the Padaung of North Thailand, swidden agriculturalists famous for wearing many neck rings to extend their necks, only married and shared their inherited territory with other Padaung and are highly identifiable as Padaung. Similar endogamy is practised by other hill tribes in the same region. And all over the world. The practice of limited exogamy preserves territory within clans and tribes. Too wide-roving exogamy ultimately means loss of control over inherited territory.

[170] Shepher, J. (1983) *Incest, A biosocial view*, Academic Press, New York.

[171] Dawkins, Richard, 1976. *The Selfish Gene*. New York City: Oxford University Press.

[172] Waldman, Bruce, Rice, John E. , Honeycutt, Rodney L., (1992) "Kin Recognition and Incest Avoidance in Toads," Integrative and comparative biology, Volume 32, Number 1, pp. 18-30.

[173] Jarne P. and Charlesworth, D. 1993. The Evolution of the selfing rate in functionally hermaphroditic plants and animals. *Annual Review of*

Ecology and Systematics. 24:441–466. 10.1146/annurev.es.24.110193.002301 and http://www.ncbi.nlm.nih.gov/pmc/articles/PMC1560239/#bib21.

[174] Engelstädter, Jan and Charlat, Sylvain, (2006 April 22) "Outbreeding selects for spiteful cytoplasmic elements," *Proceedings of the Royal Society, Biological Sciences*, 273(1589): 923–929, published online. doi: 10.1098/rspb.2005.3411, 2006, January 17.

[175] Fletcher, D. (2006) "Population Dynamics of Eastern Grey Kangaroos in Temperate Grasslands" (PHD thesis), Inst of Applied Ecology, University of Canberra. http://erl.canberra.edu.au/uploads/approved/adt-AUC20070808.152438/public/02whole.pdf: "A third approach explicitly involves more than one trophic level, by relating population growth rate to an ecological factor such as food availability, thus conforming to the 'mechanistic paradigm' (Caughley and Sinclair 1994; Krebs 1995, 2002). This is the least commonly adopted conceptual approach but it is the one underlying this study."

[176] Pimentel, D., 1968, March. Population Regulation and Genetic Feedback. *Science*, Vol.159, p. 1434.

[177] Andrewartha, H.G., and Birch, L.C. (1954) *The Distribution and Abundance of Animals*, University of Chicago Press, Chicago, cited by David Pimentel (1968, March) in "Population Regulation and Genetic Feedback," *Science*, Vol.159, 29, p.1433.

[178] Ibid.

[179] Hopfenberg, R. and Pimentel, D. (March 2001) "Human Population Numbers as a Function of Food Supply", 1 Duke University, Durham, NC, USA; 2 Cornell University, Ithaca, NY, USA http://panearth.org/WVPI/Papers/HumanPopulationNumbers.pdf

[180] Hopfenberg and Pimentel, *op. cit.*, p.4.

[181] Ibid.

[182] Hone, Jim, and Sibly, Richard M., 2002, Sep 29. "Demographic, mechanistic and density-dependent determinants of population growth rate: a case study in an avian predator," in 1: *Philos Trans R Soc Lond B Biol Sci.* 357(1425):1171-7. Applied Ecology Research Group, University of Canberra, Canberra 2602.

"The negative relationship between population density and population growth rate is at the heart of population biology. [...]   "Identifying the determinants of population growth rate is a central topic in population ecology. Three approaches (demographic, mechanistic and density-dependent) used historically to describe the determinants of population growth rate are here compared and combined for an avian predator, the barn owl (Tyto alba). The owl population remained approximately stable (r approximately 0) throughout the period from 1979 to 1991. There was no evidence of density dependence as assessed by goodness of fit to logistic population growth. The finite (lambda) and instantaneous (r) population growth rates were significantly positively related to food (field vole) availability. The demographic rates, annual adult mortality, juvenile mortality and annual fecundity were reported to be correlated with vole abundance. The best fit (R(2) = 0.82) numerical response of the owl population described a positive effect of food (field voles) and a negative additive effect of owl abundance on r. The numerical response of the barn owl population to food availability was estimated from both census and demographic data, with very similar results. Our analysis shows how the demographic and mechanistic determinants of population growth rate are linked; food availability determines demographic rates, and demographic rates determine population growth rate. The effects of food availability on population growth rate are modified by predator abundance.

Lutz, W., Testa, M.R., and Penn, Dustin J. 2006 March. Population Density is a Key Factor in Declining Human Fertility. Population & Environment, 28:69-81 DOI 10.1007/s11111-007-1137-6:

"Reproduction has been found to decline with increasing population density in a wide variety of species, yet demographers have not given systematic attention to density as a relevant factor in human reproduction".

Although density variations carry their own explanation in low environmental fertility, this self-evidence seems to escape many modern planners and economists who, seeing a sparsely settled area will advocate that it be put to more intensive use.

[183] Hopfenberg and Pimentel, 2001. *Op. cit.* p.4:

"The finding that the population size of animal species is a function of food avail-ability has been empirically demonstrated. Food energy is partitioned into four compartments viz.: maintenance, growth, stored energy,and reproduction. Scott and Fore (1995) investigated the effects of food availability onreproduction in the marbled salamander. Subjects were assigned to one of three groups. At the end of the experiment, 60% of the high-food females were reproductive. In the medium-and low-food groups, these numbers were 42% and 12% respectively. These results demonstrate that food availability influences the population dynamics of a species.

Similarly, Komdeur (1996) demonstrated that the Seychelles warbler prolonged their reproductive season, including increases to year-round breeding, when their natural condition changed to one with high food availability. Conversely, in female musk shrews (whose sexual receptivity is not restricted to the preovulatory period), 48 h of food restriction led to reduced mating behavior compared with ad-lib controls.

Thus, small reductions in food availability can inhibit female sexual behavior (Gill and Rissman, 1997). In the Calanus finmarchicus, egg production is suppressed when the nutrient pool decreases below a minimal critical value. There-after, no eggs are laid. When food is reintroduced, somatic growth resumes until structural body weight is restored, then oogenesis is fueled (Carlotti and Hirche, 1997). Also, Iwamoto (1978) has shown that monkey troop size increases rapidly after artificial provisioning, but the level of consumption efficiency of the troop is always maintained lower than the critical point in both the artificial and natural habitat. Starvation within the troop simply does not occur if the rate of food avail-ability is held relatively constant. Under natural conditions, as the feeder population increases, the food population decreases. This leads to a decrease in the feeder pop-ulation which is then followed by an increase in the food population. This increase in food availability again produces an increase in the feeder population. In quaternary consumer species, the so-called 'top of the food chain', this occurs primarily through fluctuations in birth rates."

[184] Ibid.

[185] Joseph B. Birdsell, 1971, *Op. cit.*. pp.334-361. Birdsell suggested it: "Tindale's (1940) map of the territorial boundaries of approximately 400 dialectical tribes reveals an orderly spacing of people throughout the continent, and although the intensity of land use varied dramatically with changing environments, there were no empty or unclaimed spaces.

From Tindale's basic data Birdsell (1953) produced a quantitative ecological analysis which demonstrated the importance of a single environmental variable, mean annual rainfall, in determining the size and pop density of tribal areas. A single enviro determinant cannot completely express the relationship of the numbers of men to their land, but in this case a coefficient of cuvilinear correlation of 0.18 was obtained. This close relationship was determined for basic ecological units, the 123 dialectical tribes whose resources primarily depended upon locally earned rainfall. The mathematical function of the relationship was logarithmic and areas decreased and densities increased as the mean annual rainfall rose. The analysis showed continuous variation, as opposed to stepped variation, between the minimum rainfall of four inches, through the total range to the maximum of 160 inches annually in the rain forests of north-east Queensland."

[186] Ekblom, Robert, 2000. Inbreeding avoidance through mate choice. Introductory essay. Evolutionary Biology Centre, Department of Population Biology, Uppsala University, Norbyvägen 18 D, SE-752 36 Uppsala, Sweden, http://www.ebc.uu.se/popbio/people/rekblom/Ekblom%202000.pdf

"4.2 Reproductive suppression and delayed maturity One way of dealing with inbreeding depression in animals that live in family groups, is delayed sexual maturation (or other forms of suppression of sexual activity) when the parent of the opposite sex is still present in the group (Pusey and Wolf, 1996). This phenomenon is known in prairie dogs (Cynomus ludovicianus), where a young female is significantly less likely to come into oestrus when her father is still living in her natal coterie territory (Hoogland, 1982, 1992). The mechanisms behind this are, as yet, unknown but one possibility is that hormonal activity could be affected by the scent of related individuals (Blouin and Blouin, 1988). In the communally nesting acorn woodpeckers (Melanerpes formicivourus) a female does not breed in her natal territory until her father has been replaced. Other females of the same age, however, reproduce after having migrated out from their natal group (Koenig and Pitelka, 1979). No true assessment of genetic similarity is needed for this kind of incest avoidance. Individuals must, however, be able to identify different individuals of their family and some kind of regulation of reproduction in response to this is necessary."

[187] Ekblom, Robert, 2000, *Op. cit.*: "It is important to note that mate choice and dispersal are non-exclusive to one another. Instead, one

common driving force of male dispersal is competition for mates. If females prefer to mate with unrelated males, then males should profit from dispersing out of the natal territory to find unrelated females. Such female mediated male migration has been found in for example olive baboons (Papio anubis) (Packer, 1979) and white-footed mice (Peromyscus leucopus) (Keane, 1990). In this way, female mammals could force males to bear the high cost of inbreeding avoidance (dispersal) by using a cheap mechanism (choice). In birds, females are usually the dispersing sex, a fact that has puzzled investigators of inbreeding avoidance (Pusey, 1987). If the costs of dispersal are smaller in birds than in mammals, this could possibly lead females to disperse themselves instead of forcing the males to disperse by means of mate choice."

[188] Wilson, E.O., (2001) "Nature Matters", *American Journal of Preventative Medicine*, Apr;20(3):241-2.

[189] Koenig, W., and Haydock, J. No date. Social Behavior of the Cooperatively Breeding Acorn Woodpecker. Hastings Reservation and Museum of Vertebrate Zoology. University of California Berkeley, http://www.hastingsreserve.org/AcrnPkrs/AcrnPkrs.html (Accessed 2004) Now at http://www.hastingsreserve.org/Resident%20Web%20Pages/Koenig%20Web%20Pages/AWIntroPoster/AWposter.html. (Study began in 1971) (No date). Accessed 2 November 2009. Page modified on 21 March 2012.

[190] Cockburn, A, Osmond, H, Mulder, R, Green, J, Double, C. 2003. Divorce, dispersal and incest avoidance in the cooperatively breeding superb fairy-wren Malurus cyaneus. Journal of Animal Ecology, Volume: 72, Pages: 189-202.

[191] Marie Charpentier, Patricia Peignot, Martine Hossaert-McKey, Olivier Gimenez, Joanna M. Setchell, and E. Jean Wickings., 2005. Constraints on control: factors influencing reproductive success in male mandrills (Mandrillus sphinx). CEFE-CNRS UMR 5175, 1919 Route de Mende, 34293 Montpellier Cedex 5, France, UGENET, CIRMF, Franceville, Gabon, and Department of Biological Anthropology, University of Cambridge, Cambridge, UK, *Behavioral Ecology* 16:614–623]

[192] Articles cited are: Cockburn A, Osmond HL, Mulder RA, Green DJ, Douvle MC, 2003. Divorce, dispersal and incest avoidance in the cooperatively breeding superb fairy-wren Malurus cyaneus. J Anim Ecol

72:189–202; Griffin AS, Pemberton JM, Brotherton PNM, McIlrath G, Gaynor D, Kansky R, O'Riain J, Clutton-Brock TH, 2003. A genetic analysis of breeding success in the cooperative meerkat (Suricata suricatta). Behav Ecol 14:472–480; Mateo JM, 2003. Kin recognition in ground squirrels and other rodents. J Mammal 84:1163–1181; Pusey A, Wolf M, 1996. Inbreeding avoidance in animals. Trends Ecol Evol 11:201–206; Stow AJ, Sunnucks P, 2004. Inbreeding avoidance in Cunningham's skinks (Egernia cunninghami) in natural and fragmented habitat. Mol Ecol 13:443–447; Yu XD, Sun RY, Fang JM, 2004. Effect of kinship on social behaviors in Brandt's voles (Microtus brandti). *J Ethol* 22:17–22.

[193] Charpentier, M., Peignot, P., Martine Hossaert-McKey, M., Gimenez,,O., Setchell, J.M. and Wickings, E.J., 2005, *op. cit.*

[194] Lukas, D., Reynolds, V., Boesch, C. and Vigilant, L. 2005. To what extent does living in a group mean living with kin? *Molecular Ecology* 14, 2181–2196 doi:10.1111/j.1365-294X.2005.02560.x. http://www.eva.mpg.de/primat/staff/boesch/pdf/mol_eco_what_ext.pd f. Actually the authors claim that this female dispersal is the rule in human groups but I would suggest that this is not an invariable.

[195] Dallwig, Rebecca, 2008, September 11. The Common Marmoset, Callithrix jacchus, Current Research. Page 8, III. Biomedical Research, in Primate Info Net, Library and Information Service, National Primate Research Center, University of Wisconsin – Madison, http://pin.primate.wisc.edu/callicam/research8.html. Site records indicate last modification to site was September 11, 2008. (Last accessed 21 March 2012) This website reference contains summaries from research described more fully in the following documents: Abbott, D. H., Saltzman, W., Schultz-Darken, N. J., & Tannenbaum, P. L. (1998); Abbott, D. H.,Saltzman, W., Schultz-Darken, N. J., & Smith, T. E. (1997); Baker, J. V., Abbott, D. H., &Saltzman, W. (1999).

[196] Dallwig, Rebecca, 2008. The Common Marmoset, Callithrix jacchus, Current Research, *op. cit.*

[197] Hoier, S., 2003. "Father absence and the age of menarch, A test of four evolutionary models," *Human Nature*, Vol. 14, No. 3, pp. 209–233, Walter de Gruyter, Inc., New York.

[198] The figure was adapted from diagrams by Brian Schwimmer. Schwimmer, Brian, 1995. Hebrew social organization. Lecture series.

Department of Anthropology, University of Manitoba.
http://umanitoba.ca/faculties/arts/anthropology/tutor/case_studies/he
brews/marriage.html#levirate. (Accessed 2 November 2009). Permission
to use diagrams has been obtained.

"These diagrams are geometric illustrations of incest avoidance
prescriptions from Leviticus 18. 'Ego' is the central reference person.
Patrilineally related men are in blue; Excluded marriage partners are in
red; Partners not explicitly excluded are in green."

[199] *Leviticus* 18, "Unlawful Sexual Relations",

1 The LORD said to Moses, 2 "Speak to the Israelites and say to
them: 'I am the LORD your God. 3 You must not do as they do in Egypt,
where you used to live, and you must not do as they do in the land of
Canaan, where I am bringing you. Do not follow their practices. 4 You
must obey my laws and be careful to follow my decrees. I am the LORD
your God. 5 Keep my decrees and laws, for whoever obeys them will live
by them. I am the LORD.

6 " 'No one is to approach any close relative to have sexual relations.
I am the LORD.

7 " 'Do not dishonor your father by having sexual relations with
your mother. She is your mother; do not have relations with her.

8 " 'Do not have sexual relations with your father's wife; that would
dishonor your father.

9 " 'Do not have sexual relations with your sister, either your father's
daughter or your mother's daughter, whether she was born in the same
home or elsewhere.

10 " 'Do not have sexual relations with your son's daughter or your
daughter's daughter; that would dishonor you.

11 " 'Do not have sexual relations with the daughter of your father's
wife, born to your father; she is your sister.

12 " 'Do not have sexual relations with your father's sister; she is
your father's close relative.

13 " 'Do not have sexual relations with your mother's sister, because
she is your mother's close relative.

14 " 'Do not dishonor your father's brother by approaching his wife
to have sexual relations; she is your aunt.

15 " 'Do not have sexual relations with your daughter-in-law. She is
your son's wife; do not have relations with her.

16 " 'Do not have sexual relations with your brother's wife; that would dishonor your brother.

17 " 'Do not have sexual relations with both a woman and her daughter. Do not have sexual relations with either her son's daughter or her daughter's daughter; they are her close relatives. That is wickedness.

18 " 'Do not take your wife's sister as a rival wife and have sexual relations with her while your wife is living."

[200] The same applies to disturbed populations of other species. For instance, in kangaroos, which are frequently and violently 'culled' to 'prevent' kangaroo overpopulation in Australia, e.g. the Belconnen Kangaroo case, where 500 out of 600 kangaroos were killed because it was anticipated that they would overshoot the capacity of the artificially created island they were confined to in a sea of farms and new housing estates in Canberra. The rest were implanted with contraceptives or forcibly sterilized. It was claimed that the kangaroos had tripled their population over the preceding two years. If this were true I surmised that, either the population had grown due to immigration (which had not been considered), or it was an artifact of immigration, and had been artificially formed by dislocation (due to human expansion) of kangaroos all over the state, giving a starting population of highly genetically disparate kangaroos, without the benefit of any Westermarck effect. Had the 600 healthy kangaroos been made up of intact family and clan structures then their fertility would have been curbed by incest avoidance and the Westermarck effect. Unfortunately for these benighted animals, the solution undertaken by the humans who were trying to 'manage' the population, was to drive them randomly into a cul de sac and shoot them. There was no consideration, for humane purposes, and certainly not for Westermarck effect enhancement, of preserving some families intact. And there was no consideration, apparently, that more kangaroos might migrate into the area, since it was not completely impermeable. So, by killing 500 kangaroos indiscriminately, space was liberated for new kangaroo-immigrants from clans all around with no inhibitions about mating with the surviving 100 kangaroos and with each other. By fracturing the clan structures any barriers to fertility that were already present had been removed and space had been created for 500 new and probably unrelated kangaroos. That represents a potential explosion of fertility, even if the contraceptives work. In the mean-time, the kangaroos

that suffered the cull have lost the order, protection, security and affection that comes from being part of a clan. Not what you would call humanely or scientifically a win-win solution. Sheila Newman, 2009, May 25. ACT Roo killings: Who profits? Behind the Earless Dragon mask. http://candobetter..net/node/1274

[201] Mi-Kyung Cho, "Korea: The 1990 Family Law Reform and the improvement of the status of women," 33 U. Louisville *J. Fam. L.* 431 1994-1995. Regarding Central Australian Aborigines see Joseph B. Birdsell, 1971, *op. cit.*

[202] This initial explanation does not go into detail of variation in the way that families and societies organise. As mentioned earlier in this book, it was, for instance, quite common for women to have their own villages, where young children of both sexes are raised until the age of about six, when the male children are sent to be raised in the men's villages. This situation prevails in some more traditional indigenous societies, especially in remote Pacific Islands and parts of New Guinea.

[203] Abernethy, 2005, *op. cit.* pp 64-65.

[204] Levi-Strausse, Claude, 1967. Les structures élémentaires de la parenté (Elementary Structures of Kinship). Mouton et Maison des sciences de l'Homme, Paris, La Haye. (First edition 1947.) On the oppositional structure of incest avoidance, see earlier section this book.

[205] A topology is a model, a mould, a shape (or a system) that may vary but which is always recognisable by the retention of some essential characteristics, without which it will not work. The term comes from geometry.

[206] "Kinship Relations" in "Share our Pride," http://www.shareourpride.org.au/topics/culture/kinship-protocols. (Accessed 31 August 2012.)

[207] This was a term used by Émile Durkheim, 1967 (1893), *op. cit.*, to denote societies which were held together 'mechanically' by relatively rigid laws of social conduct and dress, in contrast to 'contractual' societies where laws do not tend to be so personal, with wide variations in dress and social behaviour, negotiated according to circumstances.

[208] This is the predominant hypothesis explored in Chepko-Sade, B.D., and Tang Halpin, Z., 1987. *Mamalian Dispersal Patterns*, Chicago Press, Chicago.

[209] Hoier, S. 2003, *op. cit.*

[210] Although the Westermarck Effect is never canvassed as such here, a case of a stallion mating with his daughter and of a wolf 'step-father' (who moved in with a female and her cubs after the blood father died) forming a union with the mother's daughter, with the mother leaving the homesite, are described in B. Diane Chepko-Sade, B.D. and Tang Halpin, 1987, *op. cit.*, p. 50 (in horses) and p.59 (in wolves.) In these cases there had been no opportunity for a Westermarck effect to form.

[211] Nwe, Than Than, 2005. Aboriginal cultures and the land: An introduction to the unit. Course at Central Queensland University (Australia), www.humanities.cqu.edu.au/abtorres/52246/52246sg., p.73.

[212] Nwe, Than Than, 2005, *op. cit.*, 1–15, p.21.

[213] (See also Schwimmer, Brian, 1995, *op. cit.* on "Parallel Cousin Marriage and Lineage Endogamy" in the notes below.) Such as with brother/sister among Egyptian pharaohs, or in ancient Greek settler societies in African colonies, and among first cousins in Karen [swidden farming, North Thailand] society, to go by tradition elicited in Ferguson, F., 2007. *Look Down, see the women cry*, from a Karen folktale he recorded as "7. The story of Nauj Htof K'Maiz and Cau Seif Laf Geiz, (Taj Leplez Nauj Htof K'Maiz dauv Cau Seif Laf Geiz,)" available in Folktalkes/tajleplez.pdf, archived at http://tonbo80.spaces.live.com/.

[214] Schwimmer, Brian, 1995, *op. cit.*, "The levirate is a widespread institution, which requires that a man becomes the husband of a deceased brother's widow. In the biblical text this imposition is seemingly restricted to a situation in which both brothers reside in the same household and where the deceased has no son to succeed him. It is justified in terms of the need for him to have an heir so that "his name may not be blotted out of Israel(Deuteronomy 25:5)". In this regard, the dead brother rather than the living biological parent becomes the acknowledged or "sociological father" of the child, especially in regard to the establishment of an official genealogical line." (See the story of Judah and Tamara (Genesis 38)

[215] Schwimmer, Brian, 1995, *op. cit.*, "Parallel Cousin Marriage and Lineage Endogamy

The second substantial prescription is also related to the manipulation of marriage ties in order to ensure continuity within the lineage on occasions in which a man has only daughters. In this case, the daughters inherit his property but are married off to their patrilateral parallel cousins (their father's brother's sons)(Numbers 36). This mechanism allows the property and line of descent to remain within the patrilineage since a daughter's husband belongs to the same lineage as his wife and the children are placed within the patriline through both parents. Accordingly this form of marriage is also referred to as lineage endogamy, i.e., marriage within the lineage.(See Sagas of the Hebrew Patriarchs for a detailed illustration of endogamous lineage development the Hebrew origin myth.)"

[216] Schwimmer, Brian, 1995, *op. cit.*, Hebrew social organisation : Marriage.

[217] As with other illustrations, I have modified it for black and white reproduction and changed the size and lay-out to suit the publishing medium.

[218] This diagram is adapted from an illustration on-line by Brian Schwimmer. Brian, 1995. Hebrew social organization. Lecture series. Department of Anthropology, University of Manitoba. http://umanitoba.ca/faculties/arts/anthropology/tutor/case_studies/he brews/marriage.html (Accessed 2 November 2009). Permission to use diagrams has been obtained.

[219] Mowat, F. 1963. *Never Cry Wolf.* McClelland and Stewart and B. Diane Chepko-Sade and Zuleyma Tang Halpin, 1987, *op. cit.*

[220] Hopfenberg and Pimentel, 2001, March, *op. cit.*

[221] As I edit Book One for publishing, it has so far taken eight years. Books 2 and 3 are written, but not yet edited for publication, and Book 4 is not yet finished.

[222] This subject is treated in detail in Books Two and Three.

[223] For more on this see Newman, Sheila, 2008. "France and Australia after oil," in *The Final Energy Crisis*, second edition, Pluto Press, UK, pp241-260.

[224] Wilson, E.O., 2001, *op. cit.* "Incest avoidance responses are recognized to occur between people who are raised in close proximity. This has been demonstrated in a 'kibbutz effect' where children raised in the same Israeli kibitzes were not attracted to each other. Several disciplines, prominently including biological anthropology, sociobiology, cognitive psychology, and neuroscience, are yielding evidence that other

innate algorithms affect the development of human behavior. These algorithms can be blocked or reversed only at the peril of mental health. An example is the negative imprinting that forms the basis of incest avoidance, as follows: When either of two persons lives in close domestic proximity during the first 30 months' life of either one, both are unable to form close sexual bonding later in their lives. The phenomenon, known as the Westermarck effect in honor of the Finnish anthropologist who discovered it a century ago, is evidently widespread if not universal in human beings. Equally impressive, it is shared by all other primate species whose sexual behavioral development has been closely studied."

[225] Sinnamon, James, 2009, 24 January. "How the Growth Lobby threatens Australia's future, http://candobetter.net/node/1002

[226] Tainter, Joseph, 1988. *The Collapse of Complex Societies*, Cambridge University Press, gives the example of Europe, where total collapse has not occurred for a long time.

[227] Birdsell, Joseph B., 1971, *op. cit.* Australia: Ecology, spacing mechanisms and adaptive behaviour in aboriginal land tenure, in Ron Crocombe, (Ed.), *Land Tenure in the Pacific*, OUP/MUP 1971, pp.334-361.

[228] Australian Aborigines come in a wide variety of regional clans and do not all have the same origins, even though those origins are all fairly ancient.

[229] Gray, Martin, 2005, The Moai statues of Rapa Nui. Sacred Sites. (The site says Copyright 2010, but contains a 2005 publication.) http://www.sacredsites.com/americas/chile/easter_island.html : "Located in the Pacific Ocean at 27 degrees south of the equator and some 2200 miles (3600 kilometers) off the coast of Chile, it is considered to be the world's most remote inhabited island. Sixty-three square miles in size and with three extinct volcanoes (the tallest rising to 1674 feet), the island is, technically speaking, a single massive volcano rising over ten thousand feet from the Pacific Ocean floor. The oldest known traditional name of the island is Te Pito o Te Henua, meaning 'The Center (or Navel) of the World.' In the 1860's Tahitian sailors gave the island the name Rapa Nui, meaning 'Great Rapa,' due to its resemblance to another island in Polynesia called Rapa Iti, meaning 'Little Rapa'. The island received its most well known current name, Easter Island, from the Dutch sea captain Jacob Roggeveen who became the first European to visit Easter Sunday, April 5, 1722."

[230] *Ibid.* The settlers, numbering less than 100, arrived on the 117 sq.km island in approximately 390 or 400 AD. Their civilisation, although tiny, had writing and monuments. It lasted less than 900 years but longer than many civilisations we currently consider important. The leader of the expedition divided the land up between his sons. By 800 AD the forest on the island was already severely depleted. At peak monument production stage, between 1000 AD and before 1500 AD, the population was around 7000, organised into 15 main clans, with land allocated in usual pacific islander pie slice style. The clans' religious, political and social centres were located on the coastal area of their slices. In about 1500 AD the population rose to its highest, around 10,000, crashing soon after. No-one is sure of why. By 1700 AD population dropped to about one tenth or one quarter of its former number – around 2,200 or 1000 people then. In 1722 first reported European contact.

[231] Diamond, J. 2005, *Collapse*, Viking Penguin/Allen Lane, New York and London, p116. "What affects Deforestation on Pacific Islands? Deforestation is more severe on: dry islands than wet islands; cold high-latitude islands than warm equatorial islands, old volcanic islands than young volcanic islands; islands without aerial ash fallout than islands with it; islands far from Central Asia's dust plume than islands near it; islands without makatea than islands with it; low islands than high islands; remote islands than islands with near neighbors; and small islands than big islands."

[232] Hunt, T.L., 2007, Rethinking Easter Island's ecological catastrophe, *Journal of Archaeological Science* 34:485-502, p.493

[233] My source for this information about mitochondrial DNA was a one off correspondence with Erika Hagelberg, molecular biologist, University of Oslo, http://www.cees.no/?option=com_staff&person=erikaha. Unfortunately I have lost the record of this correspondence, but it turns out, more fortunately, as my research continued, that it was not crucial.

[234] Author unknown. No Date. "Frequently asked questions", Easter Island Foundation, Easter Island Heritage website at www.Islandheritage.org/faq.html. Name and site now changed to Easter Island Foundation, http://islandheritage.org/wordpress/ which no longer provides open access to the citations originally sourced there, including those to the [Canadian] Medical Expedition to Easter Island of 1964,

listed in this list of references under "Skoryna, Stanley C." The references to the Canadian Medical Expedition to Easter Island of 1964 are also not available on the new site. The Canadian Medical Expedition to Easter Island (M.E.T.E.I.) under the direction of Stanley C. Skoryna of McGill University, was organized to study the relative role of environment and hereditary factors on an isolated population. The twenty-five person expedition included two UBC participants, Ian Efford (Zoology) and graduate student Jack Mathias. Sponsored by the World Health Organization the M.E.T.E.I had three objectives: 1) to conduct multi-disciplinary studies of native populations; 2) to study methods of collection and preservation of biological materials on field conditions, and; 3) to assist the population with medical problems. The expedition operated between October 1964 and February 1965. There are some incomplete online sources of material from this expedition, including www.alsindependence.com/Historical_Background_of_Easter_Island.htm ; www.alsindependence.com/rapanui.htm; http://www.alsindependence.com/Preliminary_Report_METEI.htm. Skoryna, Stanley G., Preliminary Report Medical Expedition to Easter Island and H.M.C.S. Cape Scott. (November 16, 1964 – March 17, 1965) http://www.alsindependence.com/Preliminary_Report_METEI.htm. (Last accessed on 20 March 2012.) (This report is sometimes referred to as, "The Canadian Expedition.")

[235] For example: Hassell, Thomas M, Harris, Emily L, (1995) "Genetic influences in caries and periodontal diseases," *Critical Review of Oral Biological Medicine*, 6(4)-.319-342 http://crobm.iadrjournals.org/cgi/reprint/6/4/319.pdf. "It is sometimes possible to approach the genetics of human disease at the population level. Studies of groupsof individuals with differing levels of inbreeding—forexample, children of marriages between cousins—and, conversely, studies of outbred or hybrid populations (e.g.,offspring of interracial marriages) offer opportunities for genetic studies. Two excellent examples of this approach can be found in the 1958 study of the effects of inbreeding on the oral conditions of Japanese children, in which almost 7,000 children were examined (Schull and Neel, 1965), and the orodental examinations of almost 10,000 children in Hawaii by Chung and co-workers (1970, 1971, 1977a,b) and Chung and Niswander (1975). In the Japanese endeavor, significant consanguinity effects were detected

for occlusal defects, enamel hypoplasia, and gingivitis, which suggests that recessive genes play a role in these conditions. The findings for gingivitis were especially indicative of a greater susceptibility to infection in the children of inbreeding, due either (1) to greater sensitivity to bacterial substances, (2) to an oral milieu more favorable to bacterial colonization or proliferation, or (3) to impairment of the host-defense mechanisms of the inbred child. In the extensive Hawaiian studies by Chung et al. (1970, 1971, 1977a,b), gingivitis was scored on anterior teeth, and an index of oral hygiene was also used. Even after the effects of oral hygiene (and some other environmental variables as well) were factored out, a significant association was found between gingivitis and racial intermixture. As shown in Table 2 (adapted from Niswander's 1975 depiction of the Hawaiian data), the children's gingivitis scores correlated positively with the score of the parental race having the lower mean periodontal index score. Niswander states that this would indicate the participation of recessive genes in increasing risk of periodontal disease (Niswander, 1975)."

[236] Author unknown. No Date. "Frequently asked questions", Easter Island Foundation, Easter Island Heritage website at www. Islandheritage.org/faq.html, *op.cit.*

[237] Fischer, Steven, R., Rongorong: The Easter Island Script: History, Traditions, Texts, Oxford University Press, 1997, pp.365-366. "The most recent analysis of mitochondrial DNA from skeletons of precontact Easter Islanders has revealed no trace of a South American genetic contact; the skeletons display exclusively Polynesian genetic profiles (Hagelberg et al., 1994: 25-6). If indicative of the historical truth, this would mean that the first people to settle Easter Island were Polynesians whose descendents did not mix with any other people during their isolation, which lasted perhaps as long as 1,700 years. It would confirm what France's leading Oceanist in the nineteenth century, Pierre Lesson, (1881:306), wrote over a century ago: "nothing authorises the belief in a race [on Easter Island] that was different and pre-existent to the arrival of the Polynesians". This argues for the settlement hypothesis of a single early colonisation from Polynesia, the model that most, but certainly not all, scientists today accept. The model was perhaps most concisely described by Lee (1986: 250): "From the beginning, both the culture and

the settlers of Easter Island were Polynesian." Where did the single canoeful of Polynesians sail from?..."

"...they were probably the first among the East Polynesian assemblage to abandon whatever island they may have been occupying in the Marquesan group homeland. Though Rapanui's archaeological record cannot reliably be dated to any earlier than AD 690 plus or minus 130 (uncalibrated; Ayres, 1973) – which involves the first construction phase at Tahai on the west coast by Hangaroa – 'since these platforms were already large and stylised, the first settlers probably arrived long before, sometime during the first centuries AD" (Bahn, 1993: 53). Linguistics can perhaps argue a more specific relative date. The Hawaiians, who shared the many linguistic innovations that characterise East Polynesian as a separate subgroup of languages, could have arrived in the Hawaiian Islands as early as AD 375 (Tuggle, 1979:189) or even AD 300 (Kirch, 1984:77). But Rapanui did not share these sundry linguistic innovations, though for many reasons the Rapanui doubtless figured amont the single original colony in East Polynesia, possibly somewhere in the North Marquesas, that "remained a coherent and isolated community for several centuries" (Clark, 1979: 259). This linguistic non-participation by the Rapanui can only be explained if the various linguistic innovations were adopted by all those descent group members who stayed behind. And with the evidence of Hawaiian, this had to be several centuries before AD 375/300."

[238] Diamond, J. *Collapse*, Viking, Penguin, USA, 2005.

[239] Tim Flannery, *The Future Eaters*, Reed Books, Australia, 1994.

[240] González-Martín, Antonio; García-Moro, Clara; Hernández, Miguel; Moral,Pedro, 2006 Mar. Inbreeding and surnames: a projection into Easter Island's past. *American journal of physical anthropology*. 129(3): 435-445. Abstract: "The population of Easter Island is one of the most interesting extant human communities due to its unique demographic history, its geographic isolation, and the development of an incomparable culture characterized by the towering "Moais" and its enigmatic writing. Following the colonization of its population by Polynesians from the Mangarevan Islands in the 5th century AD, the island remained isolated up until the middle of the 20th century. Under these conditions, with endogamy levels fluctuating between 61.04-96.54% and given such a small population, a high rate of inbreeding, and consequently, an elevated level

of genetic relationships would be expected. Using data from church and civil records, we calculated the consanguinity of the population of Rapa Nui. The results of this analysis do not support the hypothesis of a high level of consanguinity (alpha = 0.00028 and $F(t) = 0.0007$, with $F(r) = 0.00586$ and $F(n) = -0.00519$), suggesting instead the existence of a strategy used to avoid marriage between close relatives. To explain these observations, the structure and exchange dynamics of the population were studied in the tribes, known locally as "Mata." The results of this analysis suggest a tendency toward the avoidance of inbreeding within tribes, in order to decrease the rate of endogamy in each group. This is consistent with ethnographic observations from the beginning of the 20th century that support the existence of strict regulations to prevent inbreeding between closely related individuals. Furthermore, we confirm that this situation dates back to a period before the "refounding" of Easter Island. Our results demonstrate that conditions of geographical isolation are not in themselves sufficient to produce an elevated inbreeding coefficient, revealing Easter Island as an interesting example of how cultural rules can shape the genetic structure of a population."

[241] Antonio González-Martín et al, *Ibid*, p.442, citing Routledge, K.S., "The mystery of Easter Island, Hazel, Watson and Viney, London, 1919; Wilhelm, O., and Sandoval, L., 1956. Genealogia y seroantropologia de los pascuenses. *Bol Soc Biol Concep* 31:119-139; Cruz-Coke, R.,1989. Los genes del pueblo Pascuense. *Rev Med Chil* 117-685-694; and Cruz-Coke, R., Iglesisas, R., 1964. Frecuencia alelelos T,t en población de la isla de Pascua. *Arch Biol Med Exp* (Santiago) 1:29-37.

[242] Antonio González-Martín et al, *Ibid*, p.435.

[243] Anderson, A., "Faunal collapse, landscape change and settlement history in Remote Oceania," *World Archaeology* 33(3), 375-390, cited in Benny Pieser, 2005b. "From Genocide to Ecocide: the Rape of Rapa Nui", Published in: Energy & Environment, 16:3&4, pp. 513-539 and also at http://www.staff.livjm.ac.uk/spsbpeis/EE%2016-34_Peiser.pdf

[244] Hooker, Brian, 1989. Identifying Davis's Land in Maps. *Terrae Incognitae*, Vol. 21, pp. 55-61. Available on internet at Hooker, Brian, "Identifying Davis Land in maps," http://findingnz.co.nz/wc/gwc1.htm and http://nathaniel1461.tripod.com/contents.htm. There was considerable official, pirate, privateer, and buccaneer activity around South America between 1523 and the mid-1720s, notably in the Caribbean, but quite a lot

also occurred on the Pacific seaboard of South America. The buccaneers were serious traders competing with serious slavers doing official trade and exploration and there are statements that something like 40% of pirate crews were ex-slaves. We obviously don't have records of all their voyages, but, although islands are much sparser on this seaboard than on the Eastern one, they still exist and it seems likely that they were used by pirates. There is an interesting unsolved mystery about a mysterious island in the area which was described by a buccaneer called Davis. Hooker writes about the various discussions in history about an entry on a map first published in July 1714 by Guillaume de Lisle, "Hemisphere Meridional pour voir plus distinctment les Terres Australe," Paris, July, (probably) 1741. See this map on line at http://www.antarctic-circle.org/tooley.htm, where it is called, "1741 Covens and Mortier Map of the Southern Hemisphere (South Pole, Antarctic)." The entry says, "terre découverte par Davis en 1687 par la latitude de 27.0," and it looks as if it is situated just above Easter Island. Many have wondered, of course, if it was Easter Island. However others have concluded that Davis probably made a slip of the pen and wrote 'leagues' instead of 'miles' and that he may have been looking from the Isla San Felix and Isla San Ambrosio, which are islands at the coast of Chile. Brian Hooker writes: "In 1825, Frederick Beechey, attempting a solution of the "Davis's Land" puzzle, sailed in HMS Blossom to the island of Sala-y-Gómez located in 26° 28' S, 102 8' W, about 2,700 kilometers west of Isla San Felix and about 35 degrees west of Copiapo. The island bears the name of the Spanish commander who discovered it in 1793.(fn.13) Beechey's journal includes interesting commonly mistaken for land. (fn.19)." The point is that it is not impossible that Easter Island was frequented prior to its official discovery.

[245] Pieser, Benny, 2005b. "From Genocide to Ecocide: the Rape of Rapa Nui", Published in: Energy & Environment, 16:3&4, pp. 513-539 and also at http://www.staff.livjm.ac.uk/spsbpeis/EE%2016-34_Peiser.pdf

[246] Pieser, Benny, 2005b. *Ibid.* He asks how the official discoverers of Easter Island, Cornelis Bouman, Captain of Roggeveen, who wrote that "of yams, bananas and small cocnut palms, we saw little and no other trees or crops"... "no thick timber, no strong ropes" – could have known that the island was devoid of trees – over the few days they spent there.

The island is 163.6 km2 in area. And he points to the completely contrary report from Carl Frieddrich Behrens, Roggeveen's officer on the same voyage, who wrote that the islanders presented "palm branches as peace offerings" and that their houses were "set up on wooden stakes, daubed over with luting and covered with palm leaves." Behrens described the island as a "suitable and convenient place at which to obtain refreshment, as all the country is under cultivation and we saw in the distance whole tracts of woodland."

[247] Peiser, Benny, 2005. *Energy & Environment*, 16:3&4 pp. 513-539, http://www.staff.livjm.ac.uk/spsbpeis/EE%2016-34_Peiser.pdf p.519. "For a start, it remains unclear when exactly the last palm trees became extinct. Nobody questions that smaller trees existed on Easter Island up until the 20th century. There are even reports by European visitors, such as the testimony by J.L. Palmer (1870a) who claims to have spotted "boles of large palm trees" as late as the second half of the 19th century – an observation confirmed by his co-visitor Lt Dundas who also saw "a few stumps of cocoa-nut palm" (Dundas, 1871). Given these and many other uncertainties, even Flenley himself wonders whether the palm may not have vanished until much later than generally thought: "Why did the palm become extinct?

Possibly the coup de grace was administered by the sheep and goats introduced in the 19th and 20th centuries; but the species had clearly become rare before then, if Cook and La Pérouse are correct" (Flenley, 1993:35)."

[248] Flenley, J. and Bahn, P., 2003. *The Enigmas of Easter Island*, Oxford University Press, Oxford, 2003, p.123, cited by Benny Pieser, 2005b. "Whatever the case, deforestation was by no means an all-inclusive process. The smaller but important toromiro tree (Sophora toromiro) did not become extinct until the 20th century. It was essentially the only source of wood left to the islanders. Such trees provided the wood needed for housing, the building of small canoes, the carving of wooden figurines and other wooden tools and weapons. Many researchers are inclined to believe that wooden sledges or rollers produced from the toromiro tree also served as the apparatus for the statues' transportation. "The wood of the toromiro would have been suitable for rollers of 50 cm (20 in.) diameter, and also for levers, which were probably crucial to handling the statues."

[249] Peiser, Benny, 2005. *Energy & Environment*, 16:3&4 pp. 513-539, http://www.staff.livjm.ac.uk/spsbpeis/EE%2016-34_Peiser.pdf p.527. "Lavachery divided the cultural history of statue production into five periods, the last of which corresponded with the disaster brought on by European slave-raids and the natives' subsequent near-extinction. He proposed that the carving of statues in the quarries actually continued until the sculptors and their customers were taken captive and hauled off from the island by whalers and slave-raiders in the 19th century (Lavachery, 1935). In short: "For a lack of orders, the sculptors did not finish the works they had begun, and as a result of the disaster that struck the island monumental sculpture disappeared" (Metraux, 1957:161). This explanation was by far the most compelling reconstruction of the history and end of Rapa Nui's statues. Not only was there no solid evidence that the statue cult had come to an end by the time of European discovery in 1722 - in fact, the statue cult was still in practise during much of the 18th century. Unfortunately, the views by Métraux and Lavachery have been largely forgotten in contemporary discussions about the possible reasons for the cessation of the statue cult."

[250] Jarad Diamond, *Collapse*, 2005, p108, cited by Peiser, Benny, 2005. *Energy & Environment*, 16:3&4, pp. 513-539, p.519. http://www.staff.livjm.ac.uk/spsbpeis/EE%2016-34_Peiser.pdf.

[251] Peiser, Benny, 2005. *Energy & Environment*, 16:3&4 (2005), pp. 513-539, p.519. http://www.staff.livjm.ac.uk/spsbpeis/EE%2016-34_Peiser.pdf

[252] Skoryna, Stanley G., 1964-1965, *op.cit.*

[253] Peiser, Benny, 2005. *Energy & Environment*, 16:3&4 (2005), pp. 513-539, p.531. http://www.staff.livjm.ac.uk/spsbpeis/EE%2016-34_Peiser.pdf "A closer examination of his claims reveals that the accusation of "cannibalism" was a European fabrication invented during a time when European whalers and raiders repeatedly attacked the island's population. The allegation first surfaced in 1845 in a report in the French journal L'univers. According to the sensationalist tabloid-style story, the young commander of a French vessel that had landed on Easter Island fortuitously "escaped being the victim of cannibals.... Mr Olliver was brought back on board; his whole body was covered with wounds. He had, on various parts of his body, the teeth marks of these cruel islanders, who had begun to eat him alive" (Fischer, 1992: 73).Most researchers

concur that this horror-story is most likely a hoax, "one of the most ridiculous yarns ever spun about the island" (Bahn, 1997), in short the fictional fantasy of mid-nineteen century European bigotry."

[254] Pieser, Benny, 2005b, p.534.

[255] There is a very active project to gather data on and conserve the statues, many of which are buried up to their necks to a depth of perhaps five meters (two of seven meters have so far been excavated). See Van Tilburg, Jo Anne, 2009-ongoing. "Easter Island Statue Project Conservation Initiative," http://www.eisp.org/. The website contains little speculative information, however. Although it says, "The dirt and detritus partially burying the statues was washed down from above and not deliberately placed there to bury, protect, or support the statues." At http://www.eisp.org/4230/, this statement is contradicted by another "Secondly, while it is clear that natural debris has washed down from above on both the interior and exterior slopes, all upright statues were certainly backfilled. This was done, I propose, in an effort to protect them, to allow them to remain standing, and to facilitate subsequent *in situ* ritual and burial uses." http://www.eisp.org/1858/

[256] Flannery, T. 1994. *The Future Eaters*, Reed Books, NSW, Australia. "Future Eaters" is the term that Australian paleontologist, Tim Flannery, gave to peoples whose lifestyles are unsustainable because they erode the ecosystem that supports their economy.

[257] I refer to the ecological metaphor where human failure to reverse ecological degradation despite the risks it entails to human health and survival is compared to the idea that a frog, if placed in slowly heating water, will not jump out to save itself because, by the time the water endangers its life, it will already have become helpless due to earlier effects of warming. It seems hard to establish whether this is literally true – and cruel to try. See http://en.wikipedia.org/wiki/Boiling_frog

[258] There were some island exceptions to these laws, but I am excluding them from my definition of Pacific island law, which is contained in the above description of its components. Note that there is variety in traditions from island to island and depending on whether Melanesian or Polynesian etc, but I am identifying a system and constructing a topology here, not recording details of variations on the system.

[259] Bolin, Anne, 2005. *Demographics and a Historical Perspective,* (PHD). http://www2.rz.hu-berlin.de/sexology/GESUND/ARCHIV/IES/FRENPOL.HTM#DEMOGRAPHICS%20AND%20A%20HISTORICAL%20PERSPECTIVE (Last accessed 10 February 2010)

[260] Rapaport, M. (Ed.) No date. Tenure. *The Pacific Islands: Environment and Society,* Excerpts from Chapter 17. www.cba.hawaii.edu/pbcp/pdf%20Pac%20studies/tenure.pdf

[261] Bolin, Anne, 2005. Demographics and a Historical Perspective, op. cit.

[262] Religions where marriage and children are not permitted, including monastic religions, obviously also fulfill a population restraint function. The Arioi was a Tahitian cult of performers of song dance and drama devoted to the god Oro. The Arioi travelled throughout the islands by large sailing canoes, staying at individual locations long enough to give performances. They performed explicitly sexual dances and had unrestricted access to local women. First missionary condemnation and then legislation was enacted banning the cult. The name survives in some dance routines. A source for this is Craig, Robert. D., 1989. *Dictionary of Polynesian Mythology,* Greenwood Publishing Group, Conneticut, USA.

[263] The Polynesian islands tended to be more caste oriented than the Melanesian ones. Incest is undoubtedly the most important taboo, and, of course, caste was inherited, so it was affected by incest taboos.

[264] Meggitt, M.J. 1965. The Lineage System of the Mae-Enga of New Guinea, Oliver & Boyd, p93ff

[265] Abernethy, V., 1979. Population Pressure and Cultural Adjustment, p44.

[266] Bolin, Anne, 2005. Demographics and a Historical Perspective, op. cit.

[267] Religions where marriage and children are not permitted, including monastic religions, obviously also fulfill a population restraint function. The Arioi was a Society Islands cult of performers of song dance and drama devoted to the god Oro. The Arioi travelled throughout the islands by large sailing canoes, staying at individual locations long enough to give performances. They performed explicitly sexual dances and had unrestricted access to local women. First missionary condemnation and

then legislation was enacted banning the cult. The name survives in some dance routines. A source for this is Craig, Robert. D., 1989. *Dictionary of Polynesian Mythology*, Greenwood Publishing Group, Conneticut, USA. Another brief source on infanticide and the Arioi is Michel Panoff, "The Society Islands" in Ron Croncomb, (Ed.) *Land Tenure in the Pacific*, pp43-59, p.46.

[268] Bolin, Anne, 2005. Demographics and a Historical Perspective, op. cit.

[269] On the importance of perceptions, see, for instance, Virginia Deane, 1979, Population Pressure and Cultural Adjustment, Transaction Publishers, New Brunswick USA and London, UK and Fertility Decline; No Mystery, Ethics in Science and Environmental Politics (ESEP), http://www.esep.de/articles/esep/2002/article1.pdf, May 24, 2002 pp.1-11

[270] Océanie Ethnographie. 1999. *Encyclopedie Universalis*, Electronic edition, CD, 16- 720 and 16-721. The Polynesians tended to have bigger populations, some of which reached thousands and tens of thousands.

[271] Johnstone, Ian, 28 April, 2012. Background in New Flags Flying, Pacific leaders remember, Part 9, Soloman Islands, Radio New Zealand International, http://www.rnzi.com/newflagsflying/solom-bg.php.

Wood, Gordon L., *The Pacific Basin*, OUP, 1930, pp. 33:

"But the migration and wonderful development of Pacific peoples are but one side of the story. Too often the environment has proved too strong. The enervating climate, the confined space, and the easy life on Pacific islands have in many cases arrested the development of these wonderful seafarers. Civilizations rose and fell in the Pacific as elsewhere. Scattered through this island world huge monuments and abandoned fields are all that remain of great peoples who had advanced very far in the arts of civilisation. And with the coming of white men to these regions the decadence of many Pacific islanders seems to have been hastened. Isolation can bring its dangers as well as its security. Depopulation is a distressing reality in most of the islands despite the fact that labour for tropical agriculture is so urgently required; and Asiatics have had to be imported in large numbers. The future development of the islands will demand much labour, but here are the facts about declining populations. In 1870 it was estimated that Polynesia contained 690,000 native people. In 1930 there were about 200,000, but 145,000

Asiatics and 37,000 whites had come in. Melanesia was computed to have three million natives, but recently numbered scarcely one million. Micronesia declined from 273,000 to less than 90,000. In fifty years two-thirds of the native population have disappeared."

[Page 34 left hand corner has been torn off] "[?to the] contrary some of the East Indies, the Philippines [...] lands now have all the population that their resources ... permit. Some island peoples are already spilling over into neighbouring lands, and in particular the vigorous migration from Japan seems to promise that the expansion of England in the nineteenth century will find a parallel in the expansion of Japan in the twentieth. Japanese ideas and ideals re affecting the Pacific from Korea to California and from Alaska to the Argentine."

[272] Rapaport, M. (Ed.) No date. Tenure. Op.Cit.

[273] Jim McAloon. 'Land ownership - Māori and land ownership', Te Ara - the Encyclopedia of New Zealand, updated 1-Mar-09, http://www.TeAra.govt.nz/en/land-ownership/1

[274] On Sunday, 16 August 2009 Easter Islanders actually closed down the one airport there in a demonstration against unsustainable immigration, including that of tourists. This was reported by France2 television news on 18 August 2000 and elsewhere. For some detail: Sheila Newman, "Easter Islanders blocked airport against immigration-fed overpopulation this year," http://candobetter.org/node/1740 There is also a link to the French news item.

The following comments about carrying capacity and tourists overshooting it at Rapanui (Easter Island) made by a local researcher on a Rapanui mailing list and may be of interest:

Aliaga, J. M. R. 2006, July 2. New Statistics. Correspondence on Re: [easterislandinfo], Easter Island E-List.

Re: [easterislandinfo] New Statistics: 2/07/2006 2:21 AM (Melbourne time) Rapanui wrote:

"Interesting numbers... According to the only carrying capacity study we have, the island would support up to 100.000 tourists a year...but under the best -perfect- conditions: evenly distributed along the year, and with the best management for the sites: control, trails, 60 fully equippedpark rangers, no free tourists going everywhere in a 4 whell drive car, no taxi drivers doing tourism but fully professional tour guides, etc, etc. Under the present conditions, the tourism carrying capacity on the island is only

20.000 tourists a year.   I see no real way to improve this in the short term. Director of our National Tourism Service said we are working on it, but he refers to a French project with the National Park to produce thousands of mako'i, koa, aito and other trees to be planted in private parcelas -with fences- since the Rapa Nui National Park is not a real protected area -in spite of his title as the first Chilean World Heritage Site (1995). Many petroglyphs have been completely destroyed by animals inside the National Park territory. Among many endangered sites, the beautiful panel with a large tuna and Make Make in Omohi is gone forever, because of the animals.   We just finished a Japan Trust Fund project thru Unesco (about U$ 600,000.00) for Preventive Maintenance and Conservation on the island. From a previous list of more than 25 sites, the Preventive Maintenance Program treated only 6 (Crematorium at Hanga Hahave, Ahu Tarakiu, a little Ahu and a canoe ramp in Hanga Te'e, Ahu Runga Va'e and Ahu Hanga Tetenga) since Ahu Tongariki needed a large intervention. Besides, the new Lab for archaeology and conservation at the Museum is a great facility.

Some details here: www.rapanuivalparaiso.cl/conser_vacion.htm

Another news from the island is the little tsunami last June 14th. A 10 m wave washed away 40 cm of soil at the foot of Ahu Hanga Maihiku... another site to be included in the Preventive Maintenance list.

Iorana korua, Jose Miguel Ramirez Aliaga, Centro de Estudios Rapa Nui, Universidad de Valparaiso

Aliaga was responding to: [easterislandinfo] New Statistics

"Last year there were 46,320 tourists who came to the island. This is an increase of 9,459 over 2004. More than 83% arrived by air; the others by cruise ships. Visitors from Chile had the highest number, followed by England and France. USA was 4th The majority came from November to February; May-june had the lowest number of tourists. 30% of the tourists were between 25 and 34 years of age. And 35% of visitors were either professionals or executives; 15.4% were students."

[275] Fingleton, J. (Ed.), 2005, June. Privatising Land in the Pacific, A defence of customary tenures.
https://www.tai.org.au/file.php?file=DP80.pdf This is a collection of articles by six authors.

[276] Williams, Thomas, (1858) *Fiji and the Fijians* (Vol.1), London, Ed. George Stringer Rowe, introduction by Fergus Clunie, published by the Fiji Museum, Suva, 1982.

REFERENCES

Abbott, D. H., Saltzman, W., Schultz-Darken, N. J., & Tannenbaum, P. L., 1998. Adaptations to Subordinate Status in Female Marmoset Monkeys. *Comparative Biochemistry and Physiology*. 119, 261-274. http://www.biology.ucr.edu/people/faculty/Saltzmanpubs/Abbott_etal_1998_CBP.pdf

Abbott, D. H.,Saltzman, W., Schultz-Darken, N. J., & Smith, T. E., 1997. Specific Neuroendocrine Mechanisms Not Involving Generalized Stress Mediate Social Regulation of Female Reproduction in Cooperatively Breeding Marmoset Monkeys in Carter, Sue C., Lederhendler, I, Izja, & Kirkpatrick, Brian (Eds.) 1999. *The Integrative Neurobiology of Affiliation*, pp.219-238, MIT Press.

Abernethy, Virginia Deane, 1979. *Population Pressure and Cultural Adjustment*. Transaction Publishers, New Brunswick USA and London, UK.

Abernethy, Virginia Deane, 1999. *Population Politics*. Transaction Publishers, New Brunswick USA and London, UK, pp46-47.

Abernethy, Virginia, 2002, May 24, Fertility Decline; No Mystery. *Ethics in Science and Environmental Politics*. (ESEP). http://www.esep.de/articles/esep/2002/article1.pdf, pp.1-11.

Abernethy, Virginia Deane, 2004, Sep 1. Not tonight, sweetie; no energy: a neo-Malthusian looks at fossil fuels and fertility. *Worldwatch*.

Agence France-Presse, Santiago, 2009, 27 October. L'île de Pâques vote pour contrôler ses visiteurs. http://www.cyberpresse.ca/voyage/nouvelles/200910/27/01-915433-lile-de-paques-vote-pour-controler-ses-visiteurs.php.

McAloon, Jim, 2009, March 1. Land ownership - Māori and land ownership. *Te Ara - the Encyclopedia of New Zealand*, updated 1-Mar-09, http://www.TeAra.govt.nz/en/land-ownership/1

Anderson, A., 2002. Faunal collapse, landscape change and settlement history in Remote Oceania. *World Archaeology* 33(3), 375-390, cited in Benny Pieser, 2005. From Genocide to Ecocide: the Rape of Rapa Nui in: *Energy & Environment*, 16:3&4, pp. 513-539 and also at http://www.staff.livjm.ac.uk/spsbpeis/EE%2016-34_Peiser.pdf

Andrewartha, H.G., and Birch, L.C., 1954. *The Distribution and Abundance of Animals*. University of Chicago Press, Chicago, cited by David Pimentel, 1968, 29 March . in "Population Regulation and Genetic Feedback," Science, Vol.159. p.1433.

Asia for Educators, 2009. Issues and Trends in China's Demographic History. Columbia University, http://afe.easia.columbia.edu/special/china_1950_population.htm

Author unknown, Frequently asked questions. Easter Island Foundation, Easter Island Heritage website at www. Islandheritage.org/faq.html. Name and site now changed to Easter Island Foundation, http://islandheritage.org/wordpress/ which no longer provides open access to the citations originally sourced there, including those to the [Canadian] Medical Expedition to Easter Island of 1964, listed in this list of references under "Skoryna, Stanley C."

Baker, J.V., Abbott, D. H., & Saltzman, W. 1999. Social Determinants of Reproductive Failure in Male Common Marmosets Housed with their Natal Family. *Animal Behaviour*, 58,501-513.

Berkes, F. & Folke, C., (Eds.) 2000. Linking Social and Ecological Systems: Management practices and social mechanisms for building resilience. Cambridge University Press.

Bichlbaum, A., Bonanno, M. & Spunkmeyer, B. 2004. *The yes men: The true story of the end of the world trade organization*. Penguin Books. London.

Birdsell, Joseph, B. 1971. Australia: Ecology, spacing mechanisms and adaptive behaviour in aboriginal land tenure, in Ron Crocombe, (Ed.), *Land Tenure in the Pacific*. OUP/MUP, pp.334-361

Bolin, Anne, 2005. *Demographics and a Historical Perspective*. (PHD) Undated, but Internet page information shows last modified 16 March 2005. French Polynesia. http://www2.rz.hu-berlin.de/sexology/GESUND/ARCHIV/IES/FRENPOL.HTM#DEM OGRAPHICS%20AND%20A%20HISTORICAL%20PERSPECTIVE. Last accessed 21 February 2012.

Brian, M. 2012, 2 June. This popular website helps Icelandic couples avoid incest. The Next Web (TNT). http://thenextweb.com/shareables/2012/02/06/this-popular-icelandic-website-that-helps-avoid-incest/

Bourcier de Carbon, Philippe. 1998, Jan-Feb. Transition ou révolution démographique? Les insuffisances et les implications de la théorie de la transition démographique. 1: les origines [Demographic transition or revolution? The weaknesses and implications of the demographic transition theory. Part 1: the origins]. *Population et Avenir*, No. 636. Paris, France.

Borges, Dain Edward. 1992. *The Family in Bahia, Brazil, 1870-1945*. Stanford University Press, California, p.125.

Buckminster Fuller, Buckminster Fuller Institute, http://www.bfi.org/ Last accessed 22 March 2012.

Burnham, Peter, 2003. *The Concise Oxford Dictionary of Politics*. Oxford University Press.

Caldwell, J.C. 2003, July. Pretransitional Population Control and Equilibrium. *Population Studies*, Vol. 57, No. 2, pp.199-215. http://links.jstor.org/sici?sici=0032-4728%28200307%2957%3A2%3C199%3APPCAE%3E2.0.CO%3B2-L.

Catton, W. 1980. Overshoot : The Ecological Basis of Revolutionary Change Urbana: University of Illinois Press.

Catton, W. and Dunlap, R., 1978. Environmental Sociology: A New Paradigm, (1978). The American Sociologist, 13:41-49.

Caughley, G. and Sinclair, A. R. E., 1994. *Wildlife Ecology and Management*. Blackwell Science: Boston, cited in Fletcher, D. 2006. *Population Dynamics of Eastern Grey Kangaroos in Temperate Grasslands*. (PHD thesis). Inst of Applied Ecology, University of Canberra, http://erl.canberra.edu.au/uploads/approved/adt-AUC20070808.152438/public/02whole.pdf

Charpentier, M., Peignot, P., Martine Hossaert-McKey, M., Gimenez,O., Setchell, J.M. and Wickings, E.J., 2005. Constraints on control: factors influencing reproductive success in male mandrills (Mandrillus sphinx). *Behavioral Ecology*, 16:614–623. doi:10.1093/beheco/ari034.

Chepko-Sade, B.D., and Tang Halpin, Z., *Mamalian Dispersal Patterns*. Chicago Press, Chicago, 1987.

Cho, Mi-Kyung, 1994-1995. Korea: The 1990 Family Law Reform and the improvement of the status of women. 33 U. Louisville J. Fam. L. 431 1994-1995.

Cleveland, C.J., Kaufmann, R.K. & Stern, D.I., Aggregation and the Role of Energy in the Economy. *Ecological Economics*, vol.32, issue 2, pp 301-17, also at http://www.bu.edu/cees/people/faculty/cutler/articles/Aggregation_role_of_energy.pdf.

Cockburn, A, Osmond, H, Mulder, R, Green, J, Double, C., 2003. Divorce, dispersal and incest avoidance in the cooperatively breeding superb fairy-wren Malurus cyaneus, in *Journal of Animal Ecology*, Volume: 72, Pages: 189-202

Cooper, George and Daws, Gavan, 2001. *Land and power in Hawaii: the Democratic years*. University of Hawai'i Press, Honolulu, USA.

Correspondence on Easter Island E-List, "Re: [easterislandinfo] New Statistics: 2/07/2006 2:21 AM (Melbourne time) from 'Rapanui'.

Correspondence with Erika Hagelberg, molecular biologist, University of Oslo, http://www.cees.no/?option=com_staff&person=erikaha

Cruz-Coke, R., 1989.Los genes del pueblo Pascuense. *Rev Med Chil* 117-685-694. Cited in González-Martín, A., García-Moro, C., Hernández, M., Mora, P., 2006. Inbreeding and surnames: a projection into Easter Island's past. *American journal of physical anthropology.* Mar;129(3): 435-445, p.442.

Cruz-Coke, R., Iglesisas, R., 1964. Frecuencia alelelos T,t en población de la isla de Pascua. *Arch Biol Med Exp* (Santiago) 1:29-37, cited in González-Martín, Antonio, García-Moro, Clara, Hernández, Miguel, Mora, Pedro. 2006. Inbreeding and surnames: a projection into Easter Island's past. *American journal of physical anthropology.* Mar;129(3): 435-445, p.442

Daadoun, Roger, 1999. Les régulations sociales de la sexualité. *Encyclopedie Universalis*, Electronic Edition.

Dallwig, Rebecca, 2009. [Summary entitled] The Common Marmoset, Callithrix jacchus, Current Research. Page 8, III. Biomedical Research, in Primate Info Net, Library and Information Service, National Primate Research Center, University of Wisconsin – Madison, http://pin.primate.wisc.edu/callicam/research8.html (last accessed 1 Nov 2009)

Darwin, C., *On the Origin of Species by Means of Natural Selection*, 1859. Cited in Pimentel, D., 1968, March 29. Population Regulation and Genetic Feedback. *Science*, Vol.159, 29 March 1968, p.1433.

Dawkins, Richard, 1976. *The Selfish Gene.* New York City: Oxford University Press.

De Lisle, Guillaume, 1741, July. (probably). Hemisphere Meridional pour voir plus distinctment les Terres Australe. (Map) Paris. See this map on line at http://www.antarctic-circle.org/tooley.htm, where it is called, "1741 Covens and Mortier Map of the Southern Hemisphere (South Pole, Antarctic)."

Dilworth, Craig, 2010. *Too Smart for our Own Good: the ecological predicament of humankind.* Cambridge University Press, New York.

Doepke, M., 2000. Growth and Fertility in the Long Run. Mimeo, University of Chicago. Available in reduced form in Doepke, M. Accounting for Fertility Decline During the Transition to Growth. *Journal of Economic Growth* 9(3), 347-383, September 2004.

Durkheim, E., 1893. La division sociale du travail. Paris.

Dyer, Colin, 1978. *Population and Society in 20th century France.* Hodder and Stoughton, Kent, UK

Ekblom, Robert, 2000. Inbreeding avoidance through mate choice. Introductory essay. Evolutionary Biology Centre, Department of Population Biology, Uppsala University, Norbyvägen 18 D, SE-752 36 Uppsala, Sweden, http://www.ebc.uu.se/popbio/people/rekblom/Ekblom%202000.pdf.

Engelstädter, Jan and Charlat, Sylvain, 2006, April 22. Outbreeding selects for spiteful cytoplasmic elements. *Proceedings of the Royal Society, Biological Sciences*, 273(1589): 923–929, Published online 2006 January 17. doi: 10.1098/rspb.2005.3411.

Encyclopedie Universalis, 1999. Electronic CD version. Océanie Ethnographie 16- 720 and 16-721.

Falsey, Marie, 1985. Comments: Spousal Disinheritance: The New York Solution – A Critique of forced share legislation. *Western New England Law Review*, Vol. 7 (1984-1985) Issue 4 (1985), Note 5.

Ferguson, F., 2008. *Look Down, see the women cry.* https://skydrive.live.com/?cid=0a586522c3d55a14#cid=0A586522C3D5 5A14&id=A586522C3D55A14!428 /

Fingleton, Jim, (Ed.) 2005, June. Privatising Land in the Pacific, A defence of customary tenures. Discussion Paper Number 80, ISSN 1322-5421, published by The Australia Institute, https://www.tai.org.au/file.php?file=DP80.pdf.

Flannery, Tim, 1994. *The Future Eaters*. Reed Books, NSW, Australia.

Flenley, J. and Bahn, P., 2003. *The Enigmas of Easter Island*. Oxford University Press, Oxford, p.123, cited in Benny Pieser, 2005. From Genocide to Ecocide: the Rape of Rapa Nui, *Energy & Environment*, 16:3&4 (2005), pp. 513-539 and also at http://www.staff.livjm.ac.uk/spsbpeis/EE%2016-34_Peiser.pdf

Fleming, David, 2006, April 9. Why nuclear power cannot be a major energy source. Published by Feasta at www.feasta.org, climate@feasta.org and www.neweconomics .org.

Fleming, David, 2008. The Lean Guide to Nuclear Energy: A Life-Cycle in Trouble. http://www.theleaneconomyconnection.net/nuclear/Nuclear.pdf

Federation of American Scientists, No date. Heavy Water Production. http://www.fas.org/nuke/intro/nuke/heavy.htm

Fischer, Steven Roger, 1997. Rongorongo, the Easter Island Script: History, Traditions, Texts. (Monograph). Oxford Studies in Anthropological Linguistics 14. Oxford and New York: Oxford University Press.

Fischer, Steven Roger, No date. Easter Island's Rongorongo Script. http://www.netaxs.com/~trance/fischer.html (Accessed 20 March 2012.)

Freire-Maia, Newton, 1952. Frequencies of Consanguineous Marriages in Brazilian Populations, *American Journal of Human Genetics*, 4:194-203.

Fletcher, D., 2006. *Population Dynamics of Eastern Grey Kangaroos in Temperate Grasslands*. PHD thesis, Inst of Applied Ecology, University of Canberra. http://erl.canberra.edu.au/uploads/approved/adt-AUC20070808.152438/public/02whole.pdf

Freese, Barbara, 2006. *Coal, a human history*. Arrow Books, Random House London.

Feuerstein, Lionel, No date. Blocage de l'île de Pacques contre les tourists. Accessed 2009, December 27. France2 archived television video at http://info.francetelevisions.fr/video-info/index-fr.php?id-video=MAM_1500000000004645_200908181520_F2.

Griffin, AS., Pemberton, JM., Brotherton, PNM., McIlrath G., Gaynor D., Kansky R., O'Riain J., Clutton-Brock TH., 2003. A genetic analysis of breeding success in the cooperative meerkat (Suricata suricatta). *Behav Ecol* 14:472–480.

George, Annie, 2008. Widowhood in India. Demography, University of Kerala, Powerpoint production, http://www.slideshare.net/guestb82083/widowhood-in-india. (Accessed 27 August 2012).

George, Henry, 1879. *Progress and Poverty*. Abridged edition. Robert Schalkenbach Foundation, 1953.

Girard, Karine, 1993. La place des Reines de France l'adelaide d'Aquitaine a Anne de Bretagne dans l'histoire et dans l'imaginaire. Memoire de maitrise d'histoire médievale, Université Nancy II, (Histoire), http://www.hertgen.com/cvk.mmai.htm.

González-Martín, Antonio, García-Moro, Clara, Hernández, Miguel, Mora, Pedro, 2006. Inbreeding and surnames: a projection into Easter Island's past. *American journal of physical anthropology*. Mar;129(3): 435-445, p.442.

Hanna, William and Barbera, Joseph, 1962-1963. *The Jetsons*. Distributed by Screen Gems (1962–1963) and Worldvision Enterprises (1985–1987)..

Hassell, Thomas M. and Harris, Emily L, 1995. Genetic influences in caries and periodontal diseases. *Critical Review of Oral Biological Medicine*, 6(4)-.319-342 http://crobm.iadrjournals.org/cgi/reprint/6/4/319.pdf.

Hazel, Watson and Viney, Wilhelm, O., Sandoval, L., 1956. Genealogia y seroantropologia de los pascuenses. *Bol Soc Biol Concep* 31:119-139.

Hicks, Neville, 1978. This sin and scandal: Australia's population debate 1891-1911. ANU Press, Canberra.

Hoier, S., 2003. Father absence and the age of menarch, A test of four evolutionary models. *Human Nature*, Vol. 14, No. 3, pp. 209–233, Walter de Gruyter, Inc., New York.

Hooker, Brian, 1989. Identifying Davis's Land in Maps. *Terrae Incognitae*, Vol. 21, pp. 55-61, available on internet at Hooker, Brian, Identifying Davis Land in maps. http://findingnz.co.nz/wc/gwc1.htm and http://nathaniel1461.tripod.com/contents.htm.

Hone, Jim, and Sibly, Richard M., 2002, September 29. Demographic, mechanistic and density-dependent determinants of population growth rate: a case study in an avian predator. In *Philos Trans R Soc Lond B Biol Sci.* 357(1425):1171-7. Applied Ecology Research Group, University of Canberra, Canberra 2602.

Hopfenberg, R. and Pimentel, D., 2001, March. Human Population Numbers as a Function of Food Supply. Pan Earth site. http://panearth.org/WVPI/Papers/HumanPopulationNumbers.pdf

Hopfenberg, R., Hopfenberg, E., and Salmony, S., 2011. The Expansion of the Classic Demographic Transition Model. Powerpoint file. Global Population Speakout, http://www.panearth.org/

Hunt, Terry L., 2006. Rethinking the fall of the Easter Island. *American Scientist* 94(5):412-19.

Irvine, Lucy, 2001. *Faraway.* Transworld Publishers.

Jarne P, Charlesworth, D., 1993. The Evolution of the selfing rate in functionally hermaphroditic plants and animals. Annu. Rev. Ecol. Syst,. 1993; 24:441–466. 10.1146/annurev.es.24.110193.002301

Johnstone, Ian, 2012, 28 April. Background in New Flags Flying, Pacific leaders remember, Part 9, Soloman Islands, Radio New Zealand International, http://www.rnzi.com/newflagsflying/solom-bg.php.

Koenig, W., and Haydock, J., 2004. Social Behavior of the Cooperatively Breeding Acorn Woodpecker. Hastings Reservation and Museum of Vertebrate Zoology, University of California Berkeley, http://www.hastingsreserve.org/AcrnPkrs/AcrnPkrs.html (Accessed 2004) Now at http://www.hastingsreserve.org/Resident%20Web%20Pages/Koenig%20Web%20Pages/AWIntroPoster/AWposter.html. (Study began in 1971) (No date). Accessed 2 November 2009

Krebs, C. J. 2006. Two paradigms of population regulation, Wildlife Research 22, 1-10, (1995) cited in Fletcher, D., *Population Dynamics of Eastern Grey Kangaroos in Temperate Grasslands.* (PHD thesis). Inst of Applied Ecology, University of Canberra. http://erl.canberra.edu.au/uploads/approved/adt-AUC20070808.152438/public/02whole.pdf (2006)

Kaplan, H., Gurven, M., Winking, J., 2009. An Evolutionary Theory of Human Lifespan: Embodied Capital and the Human Adaptive Complex. For: *Handbook of Theories of Aging.* (Editors: Bengtson, V., Silverstein, M., Putney, N., Gans, D). Springer. Pp. 39-66. Also available http://www.anth.ucsb.edu/faculty/gurven/papers/kaplanetal_ch3.pdf.

Kurzweil, Ray, 2006. The Singularity Is Near: When Humans Transcend Biology, Penguin, London.

Lambert, B., 1971. The Gilbert Islands. In Ron Croncombe (Ed.), 1971. *Land Tenure in the Pacific*, OUP.

Levi-Strausse, Claude, *Elementary Structures of Kinship*. First Edition was 1947: *Les structures élémentaires de la parenté*. Mouton et Maison des sciences de l'Homme, Paris, La Haye, 1967

Lihoreau, M. Zimmer, C. Rivault, C., 2007. Kin recognition and incest avoidance in a group-living insect. *Behavioural Ecology*, Vol.18, No.5, 2007, pp.880-887

Lukas, D., Reynolds, V., Boesch, C. and Vigilant, L., 2005. To what extent does living in a group mean living with kin? Max Planck Institute for Evolutionary Anthropology, Deutscher Platz 6,Leipzig 04103,Germany,†Oxford University,School of Anthropology,51,Banbury Road,Oxford OX2 6PE,United Kingdom, Molecular Ecology (2005)14 ,2181 –2196 doi:10.1111/j.1365-294X.2005.02560.x

Lutz, W., Testa, M.R., and Penn, Dustin J., 2007, March. Population Density is a Key Factor in Declining Human Fertility. *Population & Environment*.

Leibenstein, Harvey, 1974. An Interpretation of the Economic Theory of Fertility: Promising Path or Blind Alley? *Journal of Economic Literature*, 12 (2).

Maddison, Angus, 2000. Monitoring the World Economy, 1820-1992. OECD, Paris.

Marx, Karl, and Engels, Frederick, 1998. *The Communist Manifesto*. Penguin, New York:, 1998

Mateo, JM., 2003. Kin recognition in ground squirrels and other rodents. *J Mammal* 84:1163–1181.

Mackillop, Andrew, with Newman, Sheila, (Eds.), 2005. *The Final Energy Crisis*. Pluto UK.

McEnery, Thornton, 2011, March 18. The World's 15 Biggest Landowners. Business Insider, http://www.businessinsider.com/worlds-biggest-landowners-2011-3?op=1

Meggitt, M.J., 1965. *The Lineage System of the Mae-Enga of New Guinea*. Oliver & Boyd.

Mowat, F., 1963. *Never Cry Wolf*, McClelland and Stewart, Toronto.

Mumford, Stephen D., 1996. *The Life and Death of NSSM 200: How the Destruction of Political Will Doomed a U.S. Population Policy*. Centre for Research on Population and Security, Research Triangle Park, North Carolina.

Munro, John, 2005, September 27. Great Britain as the homeland of the industrial revolution, 1750-1815, *The Economic History of Modern Europe to 1914*. Lecture Topic No. 4, Economics 303Y1, University of Toronto, http://www.economics.utoronto.ca/ 5/

New York UN Office, Replacement Migration, Is It a Solution to Declining and Ageing Populations? http://www.adobe.com/products/acrobat/readstep.html

Nardinelli, Clark, 2008. Industrial Revolution and the Standard of Living. *The Concise Encyclopaedia of Economics*. Library of Economics and Liberty, 2nd edition.
http://www.econlib.org/library/Enc/IndustrialRevolutionandtheStandard ofLiving.html

Nazzal, Mary. 2005, April. An Environment Destroyed and International Law, p.3.
http://www.lawanddevelopment.org/docs/nauru.pdf

Newman, Sheila, 2002. The Growth Lobby in Australia and its Absence in France, The Relationship between the Property Development and Housing Industries and Immigration Policy in Australia and France. (Research Thesis). Swinburne University, Victoria, Australia,
http://adt.lib.swin.edu.au/public/adt-VSWT20060710.144805/index.html.

Newman, Sheila, 2008. Nuclear Fission Power Options. In *The Final Energy Crisis*, Second Edition, Pluto Press, UK, p.196.

Newman, S.M.,1998. Thomas Malthus and Australian thought. In *The Social Contract*, Volume 8, Number 3 (Spring 1998)
http://www.thesocialcontract.com/artman2/publish/tsc0803/article_745 .shtml.

Newman, S.M., 2009, May 25. Land and Housing prices and Land-use planning and Housing Systems in Australia and elsewhere and the Impact of Globalisation, the Internet, Trends in Natural Increase and Immigration. Victorian Sustainable Population Australia submission to the First Home Ownership Public inquiry in 2003.
http://www.pc.gov.au/__data/assets/pdf_file/0020/58115/sub153.pdf.

Newman, Sheila, 2009. ACT Roo killings: Who profits? Behind the Earless Dragon mask. http://candobetter.org/node/1274

Newman, Sheila, 2009, December 28. Easter Islanders blocked airport against immigration-fed overpopulation this year.
http://candobetter.org/node/1740.

Nwe, Than Than, 2005. Aboriginal cultures and the land: An introduction to the unit. Central Queensland University, Australia.
www.humanities.cqu.edu.au/abtorres/52246/52246sg. P. 73. Accessed 2005. This course and the URL are no longer available.

New South Wales Government, 1804. Report of the 1804 Royal Commission into the Decline in the New South Wales Birth Rate. Described in Neville Hicks, 1978. This sin and scandal: Australia's population debate 1891-1911. ANU.

Peiser, Benny, 2005a. The Moai statues of Rapa Nui.
http://www.sacredsites.com/americas/chile/easter_island.html

Peiser, Benny, 2005b. From Genocide to Ecocide: the Rape of Rapa Nui. In *Energy & Environment*. 16:3&4, pp. 513-539. Also available at http://www.staff.livjm.ac.uk/spsbpeis/EE%2016-34_Peiser.pdf

Peterson, D., 2011. *The Moral lives of Animals*, Bloomsbury Press, New York.

Peterson, D. and Wrangham, R., 1997. *Demonic Males: Apes and the Origins of Human Violence*. Houghton Mifflin Company, New York.

Pimentel, D., 1968, March 29. Population Regulation and Genetic Feedback. *Science*, Vol.159: 1432-1437, p.1433.

Pirie, Peter, 2000, June. Untangling the Myths and Realities of Fertility and Mortality in the Pacific Islands. Population Studies Program, University of Hawaii at Manoa, Honolulu, *Asia-Pacific Population Journal*.

Prioux, F., Mazuy, M., Barbieri, M. 2010. L'évolution démographique récente en France : les adultes vivent moins souvent en couple. INED, Paris, http://www.ined.fr/fichier/t_telechargement/37336/telechargement_fichier_fr_publi_pdf1_conjoncture_3_2010.pdf

Pukrop, Michael E. 1997, May. Phosphate Mining in Nauru. TED (Trade Environment Database) Case Number 412. http://www1.american.edu/ted/NAURU.htm.

Pusey, A., Wolf, M., 1996. Inbreeding avoidance in animals. *Trends Ecol Evol* 11:201–206.

Ragreb, M., 2012, September 3. *Isontonic Separation and Enrichment*, Chapter 10.

Rapaport, M., (Ed.), Tenure. In *The Pacific Islands: Environment and Society*. Excerpts from Chapter 17. http://www.cba.hawaii.edu/pbcp/pdf%20Pac%20studies/tenure.pdf

Ronsin, Francis, 1998. *La Guerre des ventres*. Seuil, Paris.

Routledge, K.S. 1919. *The mystery of Easter Island*. Sifton, Praed and Co., London.

Schwimmer, Brian, 1995. Hebrew social organization. Department of Anthropology, University of Manitoba, http://umanitoba.ca/faculties/arts/anthropology/tutor/case_studies/hebrews/marriage.html#levirate. (Accessed 2 November 2009). Permission to use diagrams has been obtained.

Shepher, J. 1983. *Incest, A biosocial view*. Academic Press, New York.

Sinnamon, James, 2009, 24 January. How the Growth Lobby threatens Australia's future, http://candobetter.net/node/1002

Skoryna, Stanley G., Preliminary Report Medical Expedition to Easter Island and H.M.C.S. Cape Scott. November 16, 1964 – March 17, 1965. http://www.alsindependence.com/Preliminary_Report_METEI.htm. (Last accessed on 20 March 2012.) (This report is sometimes referred to as, "The Canadian Expedition.")

Smil, Vaclav, 2000. *Energies*. MIT Press.

Spencer, Herbert, 1864. *Principles of Biology*. Williams and Norgate, London.

Steadman, David W., 2006. *Extinction and biogeography of Tropical Pacific Birds*. Univ. Chicago Press, Chicago, London.

Steckel, R. H. 2004. New Light on the 'Dark Ages', The Remarkably Tall Stature of Northern European Men during the Medieval Era. *Social Science History*, 28(2):211-229; DOI:10.1215/01455532-28-2-211, Abstract.

Steele, Robert D., 2005, September 25. Review of Ray Kurzweil's The Singularity. www.amazon.com/Singularity-Near-Humans-Transcend-Biology/product-reviews/0670033847/ref=pr_all_summary_cm_cr_acr_txt?ie=UTF8&showViewpoints=1

Stow, A.J., Sunnucks, P. 2004. Inbreeding avoidance in Cunningham's skinks (Egernia cunninghami) in natural and fragmented habitat. *Mol Ecol* 13:443–447.

Sykes, Brian, 2001. The Seven Daughters of Eve: The Science That Reveals Our Genetic Ancestry. W.W. Norton, New York.

Syvret, Paul, 2009, 23 May. Damn 'em all. *Courier Mail*. Brisbane, Australia.

Tainter, Joseph, 1988. *The Collapse of Complex Societies*. Cambridge University Press.

Taylor, Paul, 2001. World Energy Map, http://www.nous.org.uk/energy.map.html

Unger, J. 1943. The Inheritance Act and the Family. *The Modern Law Review*, Vol.6., Issue 4, Dec., first published online 18 Jan 2011, http://onlinelibrary.wiley.com/doi/10.1111/j.1468-2230.1943.tb02880.x/pdf.

Van Tilburg, Jo Anne, 2009-ongoing. Easter Island Statue Project Conservation Initiative, http://www.eisp.org/.

Villanueva, J.E., 2012, October 2004. Family affairs: The two faces of political dynasties, Business World on-line, Manila.
http://www.bworldonline.com/content.php?section=12&title=Family-affairs:-The-two-faces-of-political-dynasties-&id=59508.

Waldman, Bruce, Rice, John E. , Honeycutt, Rodney L., 1992. Kin Recognition and Incest Avoidance in Toads. *Integrative and comparative biology*, Volume 32, Number 1, pp. 18-30.

Wagner, Holly, 2004, 9 January. Men from Early Middle Ages were nearly as tall as Modern people. (An article citing R.H. Steckel on his theory) Columbus, Ohio, Research News, Ohio State University, http://researchnews.osu.edu/archive/medimen.htm (Last accessed 19 September 2011 on a page published in 2004.)

Wilson, E.O., 2001. Nature Matters. *American Journal of Preventative Medicine*, Apr;20(3):241-2.

Wilkinson, Richard and Pickard, Kate, (2010) *The Spirit Level, Why Equality is Better for Everyone*. Penguin.

Williams, Thomas, 1858. *Fiji and the Fijians* (Vol.1). London. George Stringer Rowe, (Ed.). Introduction by Fergus Clunie. Republished by the Fiji Museum, Suva, 1982.

Wood, Gordon L., 1930. *The Pacific Basin*, OUP.

Youngquist, Walter, 1997. *Geodestinies*. National Book Company, Portland, Oregon.

Yu, X.D, Sun, R.Y, Fang, J.M., 2004. Effect of kinship on social behaviors in Brandt's voles (Microtus brandti). J *Ethol* 22:17–22.

Zolberg, Aristide R. 1993. Are the Industrial Countries under siege? In G. Luciani, (Ed), *Migration Policies in Europe and the United States*. Kluwer Academic Publishers, Netherlands, pp.53-82.

Veblen, Thorstein, 1899. *The Theory of the Leisure Class*, http://www.gutenberg.org/files/833/833-h/833-h.htm

INDEX

www.ingramcontent.com/pod-product-compliance
Lightning Source LLC
Chambersburg PA
CBHW030004190526
45157CB00014B/424